# Lecture Notes in Physics

For information about Vols. 1–90, please contact your bookseller or Springer-Verlag.

# Lecture Notes
# in Physics

Edited by J. Ehlers, München, K. Hepp, Zürich
R. Kippenhahn, München, H. A. Weidenmüller, Heidelberg
and J. Zittartz, Köln

## 157

## Pradip Niyogi

# Integral Equation Method
# in Transonic Flow

Springer-Verlag
Berlin Heidelberg GmbH 1982

**Author**

Pradip Niyogi
Department of Mathematics, Indian Institute of Technology
Kharagpur – 721302, India

ISBN 978-3-540-11499-4     ISBN 978-3-540-39199-9 (eBook)
DOI 10.1007/978-3-540-39199-9

2153/3140-543210

# Preface

The present work is an introduction to the integral equation method originated by Oswatitsch (1950) for studying steady inviscid irrotational transonic flow past a thin symmetric profile at zero incidence. It was the first successful method for studying a typical transonic flow field containing a supersonic region embedded in an otherwise subsonic flow. Since then, the method has been extended to the unsymmetric cases in two- and three-dimensions and established as a powerful method for studying transonic aerodynamic problems. The great rapidity of convergence of the iterative schemes in the integral equation method is a particularly attractive feature of the method, compared to the finite-difference relaxation procedures. Further, the existence and uniqueness of solution of the direct problem where the supercritical free-stream Mach number and the body shape are prescribed may be studied directly from the existence and uniqueness of the relevant integral equations. The method is promising and much work still remains to be done.

The present work is an outgrowth of a course of lectures deli - vered by the author at a Summer Institute at the Department of Applied Mathematics, Indian Institute of Science, Bangalore in 1977. Parts of the work were also lectured at the Transonic Workshop at the National Aeronautical Laboratory, Bangalore, at the Indian Society for Theoretical and Applied Mechanics (ISTAM) Congress at Rourkela in 1979 and at Kharagpur in 1980 and at the First Asian Congress of Fluid Mechanics at the Indian Institute of Science, Bangalore in 1980. The author expresses his thanks to the respective Organisers for giving him an opportunity to present the work. The

author is particularly indebted to Professor J. Ehlers, Co-editor
of the series, for going through the manuscript and in particular
for suggesting the line of development presented in section 1.3.
Thanks are due to A.K. Ray for careful checking of the typescript.
It is hoped that the present work would serve as a self-contained
account of the main features of the integral equation method for
studying transonic aerodynamic problems.

                                        Pradip Niyogi

Kharagpur - 721302,
January, 1982.

# Contents

## Chapter VII : <u>Existence And Uniqueness Question</u>

## <u>Appendix</u>

# List of Symbols

## 1. Geometrical quantities

| | |
|---|---|
| $x, y, z$ | rectangular Cartesian coordinates |
| $r, \theta, \phi$ | spherical polar coordinates |
| $X, Y, Z$ | reduced coordinates |
| $S_D, S_V, S_W$ | shock surface, vortex sheet, wing surface |
| $X_s$ | shock position |
| $E_2, E_3$ | two- and three-dimensional Eucledian spaces |
| $f(x), g(x)$ | thickness and camber distributions of profiles |
| $h(x) = \tau q(x)$ | symmetric profile shape |
| $h_u, h_l$ | upper and lower parts of the profile |
| $f(x,z), g(x,z)$ | thickness and camber distributions of wings |
| $\varepsilon$ | angle of indicence |
| $F(X), G(X), \in$ | reduced thickness, camber and angle of incidence, Eq. (1.24c) |
| $\tau$ | thickness ratio |
| $T$ | reduced thickness ratio, a transonic similarity parameter |

## 2. Analytical symbols

| | |
|---|---|
| $K_2, K_3$ | two- and three-dimensional kernels with dipole singularity, Eq.(1.70c) and Eq.(1.55b) |
| $\int_{|\cdot|} \int$ | Oswatitsch principal value of an integral |
| $\int_{\circ} \int$ | principal value, singularity removed by a circle |
| $I$ | volume/surface integrals |
| $I_T, I_C$ | thickness and camber integrals, section 4.3 |
| $a, l(a)$ | parameters, Eq.(3.17) |
| $\bar{\theta}$ | contraction factor |
| $\bar{U}_0, \bar{U}_P$ | $L_2$-norms of $U_0$ and $U_P$ |
| $\| F \|$ | $L_2$-norm of the function $F$ |

$\Phi_F(\theta)$, $\phi(\theta)$     symbol of the singular operator   F

$\overline{W}(U_0, R)$     closed ball in   $L_2$-space with centre   $U_0$ and radius R


## 3. Gasdynamic symbols

u, v, w     velocity components along rectangular Cartesian axial directions

U, V, W     reduced velocity components

$\emptyset$     velocity potential, perturbation potential

$\Phi$     reduced velocity potential (Also used as the symbol of a singular operator)

c     speed of sound

$\overline{K}$ , $\mu$     Mach number functions, Eqs. (1.18) and Eq.(3.27)

K     similarity parameter of Cole, Eq.(1.27)

$\gamma$     ratio of the specific heats

b, r     parameters

m, $\beta$     parameters, Eq.(6.12a)

E     influence function

$\overline{g}$     Correction potential, also used as a velocity function (section 5.1)

$U_P$     linearized known Prandtl solution

$U_L$     linearized solution with edge correction

$\overline{\Phi}$ , $U_H$, $\overline{U}$, $\overline{V}$     harmonic functions

$U^+$, $U^-$     symmetric and antisymmetric parts of the velocity component; also denote values downstream and upstream of a shock

$M, M_\infty$ , $M^*$     local, free-stream, critical Mach numbers

$\Delta U$     difference of velocities on upper and lower profile sides

$\Gamma$     total circulation

$\overrightarrow{q}$     velocity vector

$c_p$, $C_p$     pressure coefficient, reduced pressure coefficient, Eq.(1.20d) and Eq.(2.7)

$\Psi$ , $\Psi$ $(X, \mathcal{E} ; Y, \eta)$     elementary solution of Laplaces
                                 equation (c.f. Eq.(1.33) and Eq.(1.91))

$C_i$, $A_i$                       velocity correction at the
                                 pivotal point , Eq.(4.6) and
                                 Eqs.(5.18)

suffix $\infty$                     free-stream condition

superscript *             critical condition

References appear at the end of the work in
alphabetical order of authors, arranged chronologically.
A reference has been cited in the text by quoting the
name of the author and the year of publication of the
work. Equations of each chapter have been indicated by
arabic numerals within brackets. Equations belonging
to a different chapter have been indicated by first
quoting the chapter number followed by a decimal point
and the equation number.

# Chapter I

## Introduction and Basic Equations

### 1.1  Introduction

A flow field where both subsonic and supersonic regions are
present and are significant in determining the overall character of
the flow field is known as a transonic flow field.  Such flow fields
appear in nozzles, over propellers and turbine blades, around blunt
bodies moving supersonically and near airplanes which fly close to
Mach number unity.  Thus, with the development of modern high speed
flight vehicles, study of such flow fields have become important.
Moreover, attention has been focussed in recent times, on the question
of possible drag reduction in flight at transonic speed range.

A great amount of research work has been done over the past
thirty years on the transonic profile flow problem.  In the direct
problem, the free-stream Mach number and the profile shape are prescribed and the resulting flow field, in particular, the surface pressure distribution has to be determined.  In the indirect problems,
the body shape and the free Mach number are determined as part of the
solution.

The methods for solving transonic profile flow problem may be
divided into three main categories, viz. (a) the hodograph method,
(Ferrari and Tricomi 1968, Manwell 1971), (b) the integral equation
method (Niyogi 1977), (c) the numerical methods, (Bailey 1975).  The
methods under categories (b) and (c) may be readily extended to the
three dimensional cases of flow past a thin wing at zero and non-zero
incidence.  As the title suggests, the present lecture is devoted
mainly to the recent developments in the integral equation method in
transonic flow problems, which was originated in 1950 by K.Oswatitsch
(1950) and extended by Gullstrand (1951), Spreiter and Alksne (1955)
and Zierep (1962).  It was extended and modified further in the last
decade by Nørstrud (1968), Nixon and Hancock (1974), Niyogi (1976)
and his associates.

The earlier developments of the integral equation method by
Oswatitsch, Gullstrand, Spreiter and Alksne and Zierep have been
discussed briefly in Chapter II. Only the essential steps of the method
and important features of the solution are pointed out. The later
chapters deal in some detail with the extensions and modifications of
Nørstrud, Nixon and Hancock and Niyogi and his associates, since
among the integral equation methods, these are still promising and
capable of achieving accuracy comparable to that obtained from the
numerical finite-difference methods. Salient features of the solution
are discussed, with special reference to the modern numerical methods.

## 1.2 The integral equation method

A large number of authors have used integral equation approach
for studying mathematical and physical problems. The classical
Picard's method furnishes an example, in which an integral equation
approach was used to establish existence of the solution of an initial
value problem for an ordinary differential equation. The close relationship between partial differential equations and integral equations
have been brought forth in a number of standard works on partial
differential equations, where an integrated approach to the theory
has been presented. It suffices to mention only the works of Epstein
(1962) and Yoshida (1960). The earlier works in this area, are mainly
concerned with the question of existence and uniqueness of solutions,
and mostly discuss linear partial differential equations of a single
type by the integral equation approach. The earliest work on integral
equation approach for solving a partial differential equation of mixed
elliptic-hyperbolic type is due to Tricomi (1923) who established the
existence and uniqueness of the Tricomi problem for the Tricomi equation. Incidentally, Tricomi also showed that the cannonical form of
a second order linear partial differential equation of mixed type is
the Tricomi equation, now known after him. Subsequently a number of
authors like Frankl, Germain and Bader, Agmon used integral equation
approach for establishing existence and uniqueness of problems associated with Tricomi equation (Bers 1958 ,Ferrari and Tricomi 1968).
Existence and uniqueness of a simplified model linear equation of mixed
type, was studied by Bitsadze and Lavrentiev (Bitsadze 1964) by the
integral equation approach.

Although any procedure of studying a partial differential
equation along with boundary and/or initial conditions by inverting

it into an integral equation, may be termed an integral equation
method, with reference to aerodynamic applications, the name is
reserved for the method developed by Oswatitsch (1950) and its various
extensions and modifications carried out subsequently for studying
transonic flow fields past prescribed body shapes. It is one of the
earliest successful methods for solving the direct transonic problem,
where the free-stream Mach number and the thin body shape are prescribed, and the resulting flow field is to be computed and in particular
the surface pressure distribution is to be determined. The main idea
of the method is to invert the governing gasdynamic partial differential equations along with boundary conditions into a multi-dimensional
nonlinear singular integral equation by application of Green's theorem
of potential theory, which is then solved approximately. For profile
flow problems, the integral equation is two dimensional, while for flow
past wings it is three dimensional . For the symmetric case, only one
such equation is obtained whereas for the unsymmetric lifting case a
coupled pair of nonlinear multidimensional singular integral equations
follows. Oswatitsch considered the mathematically simplest problem of
a thin symmetric profile at zero incidence. On physical considerations,
he developed a substitution to express the velocity field approximately
in terms of that on the profile axis. Using the substitution, the
governing integral equation, now known as <u>the integral equation of
Oswatitsch</u>, reducesto a one-dimensional nonlinear singular integral
equation. Oswatitsch solved the simplified one-dimensional equation
approximately by substituting assumed forms of the solution containing
a number of disposable parameters and requiring that the integral
equation be satisfied at a number of important points of the body.
Computed solutions showed all the typical supercritical transonic
features. In particular, Oswatitsch computed solutions with shock,
while in some other cases shock-free supercritical solutions were also
obtained. Later authors extended and modified the method for studying
the unsymmetric cases in two and three dimensions.

## 1.3  Basic equations and boundary conditions in differential form

In view of the fact that we are mainly concerned here with
aerodynamic applications, it appears convenient to begin with a resumé
of the relevant basic equations of gasdynamics needed for the subsequent
study. For a fuller discussion of these topics reference may be made

to Oswatitsch (1959), Serrin (1959) and Zierep (1976).

We consider flow of an inviscid non-heat conducting perfect gas
in local thermal equilibrium, without external body forces, heat or
energy supply in a well-defined connected domain of the three-dimensional space. The basic thermodynamic variables of state of such a
system are connected by an equation of state

$$p = p(\rho, s), \tag{1}$$

where  p  denotes the pressure, $\rho$  the density and  s  the specific
entropy.

The basic principles governing the flow are the principles of
conservation of mass, momentum flux and energy flux.  The principle
of conservation of mass delivers the continuity equation

$$\frac{\partial \rho}{\partial t} + \text{div}\,(\rho\,\vec{q}) = 0, \tag{2a}$$

where  $\vec{q}$  denotes the velocity vector, assuming that there are no
fluid  sources or sinks.  The principle of conservation of momentum
flux delivers the Euler's equation

$$\frac{d\vec{q}}{dt} \equiv \frac{\partial \vec{q}}{\partial t} + (\vec{q}\cdot\vec{\nabla})\,\vec{q} = -\frac{1}{\rho}\,\text{grad}\,p, \tag{2b}$$

where  $\frac{d}{dt}$  denotes the substantial derivative, given by

$$\frac{d}{dt} \equiv \frac{\partial}{\partial t} + (\vec{q}\cdot\vec{\nabla}),$$

representing differentiation following the fluid.  The conservation
of energy flux delivers

$$\frac{ds}{dt} = 0, \tag{2c}$$

stating that the specific entropy of a fluid particle remains constant
as it moves.  This was to be expected, since in view of the  assumptions made, there is no mechanism of heat transfer or conversion of
mechanical energy  into heat in the system.  However, it is to be
noted that the specific entropy may experience jumps at any shock
discontinuity, across which the differential forms of the basic equations are no longer valid, although again valid in between shocks.

We now assume that the specific entropy  s  maintains the same
constant value in the entire upstream region upto the appearance of the
first shock in the flow field.  For such is.entropic flow, the equation
of state (1) takes the simpler form

$$p / \rho^{\gamma} = \text{constant,} \tag{3a}$$

$\gamma$  denoting the ratio of the specific heats of the gas, assumed to
be a constant quantity.  Eq.(3a) may be put to the differential form

$$\frac{dp}{p} = \gamma \frac{d\rho}{\rho}. \tag{3b}$$

For isentropic flow the speed of sound  c  is defined by

$$c^2 = (\frac{\partial p}{\partial \rho})_s, \tag{3c}$$

so that    $c^2 = \gamma p / \rho$    and  Eq.(3b) may be put to the form

$$\frac{dc^2}{c^2} = (\gamma - 1) \frac{d\rho}{\rho}. \tag{3d}$$

Eliminating the pressure and density from the continuity
equation (2a) and the Euler's equation (2b) by Eqs. (3), we obtain
the basic gasdynamic equations for isentropic flow

$$\frac{dc^2}{dt} = -(\gamma - 1)c^2 \, \text{div} \, \vec{q},$$

$$\tag{4}$$

$$\frac{d\vec{q}}{dt} = - \frac{1}{\gamma-1} \, \text{grad} \, c^2,$$

which is a system of quasi-linear equations of hyperbolic type for
the unknowns  $c^2$  and  $\vec{q}$ .

Defining the vorticity vector  $\vec{\omega}$

$$\vec{\omega} = \tfrac{1}{2} \, \text{rot} \, \vec{q}, \tag{5a}$$

the system of equations (4) deliver the well known Helmholtz equation (Oswatitsch 1959) for the vorticity vector

$$\frac{d\vec{\omega}}{dt} = (\vec{\omega}.\vec{\nabla})\vec{q} - \vec{\omega}(\vec{\nabla}.\vec{q}), \qquad (5b)$$

valid for isentropic flow. The circulation $\Gamma$ round any closed curve C consisting of fluid particles is defined as

$$\Gamma = \oint_C (u\ dx + v\ dy + w\ dz), \qquad (6a)$$

where x,y,z denote rectangular Cartesian coordinate directions and $\vec{q} = (u, v, w)$. By Stokes theorem, the vorticity vector is related to the circulation

$$\Gamma = \iint_f rot\ \vec{q}\ .\ \vec{df}, \qquad (6b)$$

f denoting a surface having C as boundary curve and the vector $\vec{df}$ denotes a directed surface element of f . For isentropic flow, it follows from Euler's equation the well known Kelvin's theorem (Oswatitsch 1959), that the circulation around any fluid curve does not change with time as it moves with the fluid. This implies that if the circulation vanishes at any time, it will remain so for all time as the fluid curve moves. In aerodynamic applications, most frequently we come across the case where the free-stream condition is a uniform state, so that the circulation around all fluid curves vanish there. A flow field where

$$rot\ \vec{q} = 0, \qquad (6c)$$

everywhere, is known as an irrotational flow. According to Eq.(6b) the circulation around any closed curve vanishes in irrotational flow. It follows then from Kelvin's theorem that an isentropic flow with uniform free-stream condition must be irrotational. The same conclusion may also be drawn from the Helmholtz vortex theorem Eq.(5b).

In the present work, we consider exclusively the case of steady flow for which the flow quantities do not depend on time. Then for steady isentropic flow, the system of equations (4) reduces to

$$\vec{q}.\vec{\nabla}c^2 + (\gamma - 1)c^2\vec{\nabla}.\vec{q} = 0,$$

and

$$(\vec{q}.\vec{\nabla})\vec{q} + \frac{1}{\gamma-1}\vec{\nabla}c^2 = 0.$$

$$(7a)$$

Consequently, expanding the vector $(\vec{q}.\vec{\nabla})\,\vec{q}$ and forming scalar product with $\vec{q}$ , it follows from the second equation of (7a)

$$\vec{q}.\vec{\nabla}\left(\tfrac{1}{2}\,q^2\right) = -\frac{1}{\gamma-1}\,\vec{q}.\vec{\nabla}\,c^2 , \qquad (7b)$$

which when integrated along a streamline delivers the Bernoulli's equation

$$q^2 + \frac{2}{\gamma-1}\,c^2 = \text{constant} = q_\infty^2 + \frac{2}{\gamma-1}\,c_\infty^2 , \qquad (8a)$$

the suffix $\infty$ denoting free-stream condition. Eq.(8a) determines the sound speed in terms of the fluid speed

$$c^2 = \frac{\gamma-1}{2}\,(q_\infty^2 - q^2) + c_\infty^2 \qquad (8b)$$

which serves to calculate the fluid pressure $p$ and the density $\rho$ . Eq.(7b) and the first equation of (7a) deliver the basic gasdynamic equation for steady isentropic flow

$$c^2\,\vec{\nabla}.\,\vec{q} = \vec{q}.\vec{\nabla}\left(\tfrac{1}{2}\,q^2\right). \qquad (9a)$$

A second basic equation for such flow is obtained from the second equation of (7a) by eliminating the sound speed by Eq. (8a)

$$(\vec{q}.\vec{\nabla})\,\vec{q} = \vec{\nabla}\left(\tfrac{1}{2}\,q^2\right),$$

which on simplification may be put to the form

$$\vec{q} \times \text{rot}\,\vec{q} = 0,$$

so that the irrotationality condition Eq.(6c) is regained apart from the exceptional case where the flow velocity $\vec{q}$ is parallel to the vorticity vector $\text{rot}\,\vec{q}$ .

It should be noted that if we drop the assumption of isentropic flow, then the flow field need not be irrotational. In such a case of steady flow with constant entropy along a stream-line, it is common to make the assumption of isoenergetic flow for which the stagnation enthalpy is constant over the whole flow field. The stagnation enthalpy does not change across shocks. Then, in stead of

the irrotationality condition (6c), one obtains (Oswatitsch 1959) as
the second basic equation, the condition

$$\vec{q} \times rot \ \vec{q} = - T \ grad \ s, \tag{9b}$$

known as Crocco's vortex theorem, where  T  denotes the absolute tem-
perature, the gasdynamic equation (9a) remaining valid in this case
also.

Crocco's vortex theorem Eq.(9b) implies that a flow field with
variable entropy is necessarily rotational.

If the shock strength is constant, then the discontinuity of
entropy across it is also constant, so that the flow after the shock
is also isentropic, i.e. grad s= 0.  It then follows from Crocco's
vortex theorem Eq. (9b) that either the flow is irrotational or the
vectors $\vec{q}$ and rot $\vec{q}$ are everywhere parallel. The latter however
is impossible since at the shock wave $\vec{q}$ has a nonzero normal component, but the normal component of rot $\vec{q}$ is zero, since it is given
by the tangential derivatives of the tangential velocity components
which are continuous across a shock.

On the other hand, if a curved shock is present, the flow field
after the shock can no more be irrotational.  However, for weak shocks
the entropy change across the shock is of the third order in terms of the
the shock strength and it has been established by Guderly (1946) and
later by Cole and Messiter (1957) and Hays (1966), that neglecting
higher order quantities the irrotationality condition is valid in the
transonic range even in the flow field after the shock, and the entire
flow field may be approximately treated as isentropic.

For irrotational flow, a velocity potential  $\emptyset$  exists defined
by

$$\vec{q} = grad \ \emptyset , \tag{10}$$

so that the condition (6c) is identically satisfied and the gasdynamic
equation (9a) leads to the basic equation

$$c^2 \nabla^2 \emptyset = \vec{\nabla} \emptyset . \vec{\nabla} (\frac{1}{2} q^2), \tag{11}$$

which is quasi-linear and has the characteristic polynomial

$$(c^2)^2(c^2 - q^2) = (c_\infty^2 + \frac{\gamma-1}{2}\left[q_\infty^2 - (\vec{\nabla}\phi)^2\right])^2.$$

$$(c_\infty^2 + \frac{\gamma-1}{2}q_\infty^2 - \frac{\gamma+1}{2}(\vec{\nabla}\phi)^2).$$

Thus Eq.(11) is of elliptic type for $q < c$, parabolic for $q = c$ and hyperbolic for $q > c$.

In terms of rectangular Cartesian coordinates $x, y, z$ the gas-dynamic equation (9a) and the irrotationality condition (6c) may be expressed respectively as

$$(1 - \frac{u^2}{c^2})\frac{\partial u}{\partial x} + (1 - \frac{v^2}{c^2})\frac{\partial v}{\partial y} + (1 - \frac{w^2}{c^2})\frac{\partial w}{\partial z}$$

$$- \frac{uv}{c^2}(\frac{\partial u}{\partial y} + \frac{\partial v}{\partial x}) - \frac{vw}{c^2}(\frac{\partial w}{\partial y} + \frac{\partial v}{\partial z}) - \frac{wu}{c^2}(\frac{\partial u}{\partial z} + \frac{\partial w}{\partial x}) = 0, \qquad (12a)$$

and

$$\frac{\partial u}{\partial y} - \frac{\partial v}{\partial x} = 0, \quad \frac{\partial w}{\partial y} - \frac{\partial v}{\partial z} = 0, \quad \frac{\partial w}{\partial x} - \frac{\partial u}{\partial z} = 0, \qquad (12b)$$

where the components of the velocity vector $\vec{q}$ are $(u, v, w)$ with $q^2 = u^2 + v^2 + w^2$.

In terms of the velocity potential $\phi$, Eq.(10) delivers

$$\frac{\partial \phi}{\partial x} = u, \quad \frac{\partial \phi}{\partial y} = v, \quad \frac{\partial \phi}{\partial z} = w. \qquad (13a)$$

From Eq. (11) we obtain

$$(1 - \frac{u^2}{c^2})\frac{\partial^2 \phi}{\partial x^2} + (1 - \frac{v^2}{c^2})\frac{\partial^2 \phi}{\partial y^2} + (1 - \frac{w^2}{c^2})\frac{\partial^2 \phi}{\partial z^2}$$

$$- 2\frac{uv}{c^2}\frac{\partial^2 \phi}{\partial x\,\partial y} - 2\frac{vw}{c^2}\frac{\partial^2 \phi}{\partial y\,\partial z} - 2\frac{wu}{c^2}\frac{\partial^2 \phi}{\partial z\,\partial x} = 0. \qquad (13b)$$

In the particular case of plane flow, $\frac{\partial}{\partial z} \equiv 0$ and $w = 0$ so that Eqs. (12) and (13) take respectively the simplified forms

$$\left(1 - \frac{u^2}{c^2}\right)\frac{\partial u}{\partial x} + \left(1 - \frac{v^2}{c^2}\right)\frac{\partial v}{\partial y} - \frac{uv}{c^2}\left(\frac{\partial u}{\partial y} + \frac{\partial v}{\partial x}\right) = 0, \tag{12c}$$

$$\frac{\partial u}{\partial y} - \frac{\partial v}{\partial x} = 0, \tag{12d}$$

and

$$\left(1 - \frac{u^2}{c^2}\right)\frac{\partial^2 \phi}{\partial x^2} + \left(1 - \frac{v^2}{c^2}\right)\frac{\partial^2 \phi}{\partial y^2} - 2\frac{uv}{c^2}\frac{\partial^2 \phi}{\partial x\, \partial y} = 0. \tag{13c}$$

It is to be noted that the local sound speed varies from point to point and according to Eq.(8a), it is a function of the flow speed $q = (u^2 + v^2 + w^2)^{1/2}$. The curves (or surfaces) along which $q = c$, that is the local Mach number equals unity are known as sonic lines (or sonic surfaces). The sonic lines are not known ápriori and are determined as part of the solution. As already observed, the governing gasdynamic equation (13b) (or 13c) ) represents a quasilinear second order partial differential equation of mixed elliptic hyperbolic type. It may be considerably simplified under the assumption of small perturbation of a uniform flow, which is undertaken in the next section.

Appropriate boundary conditions are necessary for solving Eq.(13b) or (12C). At a shock surface, Rankine-Hugoniot shock conditions are to be satisfied.

The integral equation method deals with the direct problem of steady transonic flow past a thin wing or profile shape where the uniform free stream Mach number and the body shape are prescribed and the resulting flow field is to be determined. In particular, the surface pressure coefficient and the aerodynamic force and moment coefficients are of much interest.

We consider first the case of a thin unsymmetric wing at small non-zero incidence $\varepsilon$ . We choose a body fixed rectangular Cartesian coordinate system, where the x-axis is aligned with the free stream direction, the y-axis is vertically upwards and the z-axis is along the span. The body is assumed to be thin and smooth with continuously turning tangents which are inclined at small angles with the free stream direction. Let the upper and lower body surfaces be given by

$$y = \begin{cases} h_u(x,\ z) & = & f(x,\ z) + g(x,\ z) \\ -h_l(\ x,\ z) & = & f(x,\ z) - g(x,\ z), \end{cases} \tag{14}$$

so that the functions  f  and  g  give camber and thickness

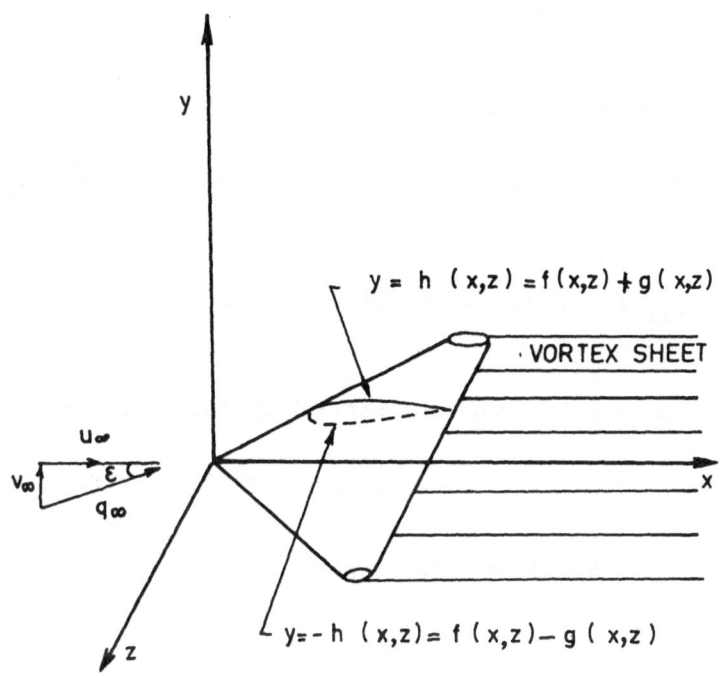

Fig.1.1    Thin unsymmetrical wing at small incidence

distributions respectively. The boundary conditions are:

(a)    the resultant flow direction at the wing surface is tangential to the surface. In other words, the normal component of the velocity relative to the wing surface vanishes at it.

(b)    The perturbation potential and its derivatives vanish at an infinite distance upstream of the airfoil. The   Kutta condition, that the pressure is finite and continuous at the trailing edge, is necessary to ensure a unique solution. For three dimensional flow, in addition to satisfying the Kutta condition, provision must be made for a trailing vortex sheet downstream of the wing trailing edge. In accordance with the small disturbance assumption, the vortex sheet is flat and lies in the plane $y = 0$, with conditions that $\emptyset_x$ (pressure) and $\emptyset_y$ (flow angle) be continuous across it. The potential $\emptyset$ and its second derivative $\emptyset_{yy}$, however, experiences a jump across the sheet.

The tangency condition (a) delivers at the body surface

$$u\, h_{ux} - v + w\, h_{uz} = 0, \text{ at the upper surface } y = h_u(x,z),$$

$$(15a)$$

and

$$u\, h_{1x} + v + w h_{1z} = 0, \text{ at the lower surface } y = -h_1(x,z),$$

$$(15b)$$

the suffixes $x$ and $z$ denoting partial differentiation.

For plane flow, the upper and lower profile surfaces are taken as

$$y = \begin{cases} h_u(x) = f(x) + g(x), \\ \\ -h_1(x) = f(x) - g(x). \end{cases} \qquad (14')$$

The corresponding tangency conditions are

$$u\, h_{ux} - v = 0 \quad \text{at the upper profile surface } y = h_u(x), \qquad (15')$$

and

$$u\, h_{1x} + v = 0 \quad \text{at the lower profile surface } y = -h_1(x).$$

The equations (15) and (15') may be considerably simplified under the assumptions of a thin smooth body producing small perturbations, which is undertaken in the next section.

## 1.4 Simplification of the basic equations and boundary conditions

Consider steady inviscid transonic flow past a thin unsymmetric body at small nonzero incidence with subsonic free stream Mach number $M_\infty$ sufficiently near unity, so that a local supersonic region embedded in a subsonic flow is formed. It is assumed that the body produces small perturbations. Then $v^2 \ll c^2$ and $w^2 \ll c^2$, so that Eq.(13) may be approximated by

$$(1 - \frac{u^2}{c^2}) \frac{\partial^2 \emptyset}{\partial x^2} + \frac{\partial^2 \emptyset}{\partial y^2} + \frac{\partial^2 \emptyset}{\partial z^2} = 0. \tag{16}$$

The approximation $w^2 \ll c^2$ is true in all cases, while $v^2 \ll c^2$ requires sharper leading edges with increasing Mach number. (This approximation fails in the hypersonic speed range, where the semi-opening angle of the body is comparable with the free stream Mach angle $\alpha_\infty$, which is however, beyond our consideration). Using Bernoulli's equation (8), the coefficient of $\frac{\partial^2 \emptyset}{\partial x^2}$ in Eq.(16) may be expanded in powers of the velocity perturbation $\frac{u - u_\infty}{u_\infty}$ as

$$1 - \frac{u^2}{c^2} = 1 - M_\infty^2 - \bar{K} \frac{u - u_\infty}{u_\infty} + \cdots , \tag{17}$$

where $\bar{K}$ is a function of $M_\infty$ given by

$$\bar{K} = M_\infty^2 \left\{ 2 + (\gamma - 1) M_\infty^2 \right\}. \tag{18}$$

According to transonic small perturbation theory, various alternative forms of $\bar{K}$ are in use. For example, Spreiter (1954) takes it as

$$\bar{K} = M_\infty^2 (\gamma + 1), \tag{19a}$$

and Oswatitsch (1950) uses the value

$$\bar{K} = (1 - M_\infty^2) \Big/ ( \frac{1}{M_\infty^*} - 1 ), \tag{19b}$$

where $M_\infty^*$ denotes the critical free-stream Mach number, defined by

$$M_\infty^* = u_\infty / c^*.$$

Let us define a reduced velocity potential $\Phi$ related to the true velocity potential $\emptyset$ by

$$\Phi(X, Y, Z) = \frac{\bar{K}}{(1 - M_\infty^2) u_\infty} \Big[ \emptyset - u_\infty x - v_\infty y \Big], \tag{20a}$$

and reduced coordinates denoted by the corresponding capital letters by

$$X = x, \quad Y = y \sqrt{1 - M_\infty^2}, \quad Z = z \sqrt{1 - M_\infty^2}. \quad (20b)$$

The reduced velocity components $U, V, W$ are then related to their true values denoted by the lower case letters by

$$\Phi_X = U = \frac{\bar{K}}{1 - M_\infty^2} \frac{u - u_\infty}{u_\infty},$$

$$\Phi_Y = V = \frac{\bar{K}}{(1 - M_\infty^2)^{3/2}} \frac{v - v_\infty}{u_\infty},$$

$$\Phi_Z = W = \frac{\bar{K}}{(1 - M_\infty^2)^{3/2}} \frac{w}{u_\infty}. \quad (20c)$$

A first order approximation to the pressure coefficient is then given by the reduced pressure coefficient $c_p$ as

$$C_p (X, Y, Z) = \frac{\bar{K}}{1 - M_\infty^2} c_p = -2U(X, Y, Z). \quad (20d)$$

For axisymmetric bodies, other terms will appear on the right side of Eq.(20d), to be discussed in the second chapter.

In terms of the reduced quantities defined above by Eq.(17) and Eq.(20) we obtain from Eq. (16) the transonic approximation to the gasdynamic equation

$$\Phi_{XX} + \Phi_{YY} + \Phi_{ZZ} = \Phi_X \Phi_{XX}. \quad (21a)$$

For plane flow, the third term on the left hand side of Eq.(21a) is to be dropped and we get

$$\Phi_{XX} + \Phi_{YY} = \Phi_X \Phi_{XX}. \quad (21b)$$

It should be noted that the reduced quantities are not small in the transonic speed range and are of the order of unity. Further, the flow is subsonic for $U < 1$ and supersonic for $U > 1$. Eq.(21) is a second order quasilinear partial differential equation of mixed elliptic-hyperbolic type, being of elliptic type in subsonic flow and of hyperbolic type in supersonic flow. Their line of demarcation, namely the sonic line (or sonic surface) is not known a'priori and must be determined as part of the solution.

Under the assumption of small perturbation, neglecting second and higher order terms, the tangency boundary condition Eqs. (15) may be shifted to the wing planform plane $y = 0$ and simplified as

$$v(x, 0, z) = \begin{cases} u_\infty \dfrac{\partial}{\partial x} h_u(x, z) , & \text{On the upper part } y = 0 + , \\[2ex] -u_\infty \dfrac{\partial}{\partial x} h_l(x, z) & \text{on the lower part } y = 0 - . \end{cases}$$

$$(22)$$

Introducing the reduced quantities by Eqs. (20) it follows

$$V(X, 0 + , Z) = \frac{\bar{K}}{(1 - M_\infty^2)^{3/2}} \left[ \frac{\partial h_u(X, Z)}{\partial X} - \frac{v_\infty}{u_\infty} \right], \quad \begin{array}{l} \text{on the part} \\ Y = 0 + , \end{array}$$

$$(23)$$

$$V(X, 0 - , Z) = - \frac{\bar{K}}{(1 - M_\infty^2)^{3/2}} \left[ \frac{\partial h_l(X, Z)}{\partial X} + \frac{v_\infty}{u_\infty} \right] \quad \begin{array}{l} \text{on the lower} \\ \text{part } Y = 0 - . \end{array}$$

The simplified form of the tangency condition for plane flow, Eq. (15') becomes in reduced coordinates

$$V(X, 0 + ) = \frac{\bar{K}}{(1 - M_\infty^2)^{3/2}} \left[ \frac{dh_u(X)}{dX} - \frac{v_\infty}{u_\infty} \right], \quad \begin{array}{l} \text{on the upper part} \\ Y = 0 + , \end{array}$$

$$V(X, 0 - ) = - \frac{\bar{K}}{(1 - M_\infty^2)^{3/2}} \left[ \frac{dh_l(X)}{dX} + \frac{v_\infty}{u_\infty} \right], \quad \begin{array}{l} \text{on the lower part} \\ Y = 0 - , \end{array}$$

$$(24a)$$

which may be rewritten as

$$V(X, 0_\mp) = F'(X) - \epsilon \pm G'(X)_1, \tag{24b}$$

where the reduced quantities are defined by

$$F(X) = \frac{\bar{K}}{(1 - M_\infty^2)^{3/2}} f(x), \quad G(X) = \frac{\bar{K}}{(1 - M_\infty^2)^{3/2}} g(x), \quad \epsilon = \frac{\bar{K}}{(1 - M_\infty^2)^{3/2}} \epsilon.$$

$$\text{(24c)}$$

In the particular case of a thin symmetric profile at zero incidence, $f(x) = 0$, $\epsilon = 0$ and $h_u(x) = -h_l(x) = g(x)$, so that the simplified boundary condition is

$$V(X, 0 \pm) = \pm G'(X). \tag{24d}$$

The boundary conditions at upstream infinity are that all perturbations vanish there and that in the downstream only the flow due to the vortex sheet leaving the wing tips is permitted. The Kutta condition is to be applied to the trailing edge: the flow leaves the trailing edge smoothly. Across the wake there can be no jump in pressure or normal velocity. For plane flow, no vortex sheet is formed and only the Kutta condition of smooth flow is to be applied at the trailing edge. The free stream condition is

$$\Phi = U = V = 0 \quad \text{as} \quad (X^2 + Y^2)^{\frac{1}{2}} \longrightarrow \infty. \tag{25}$$

An alternative form of the transonic small perturbation equations frequently used is given by Cole (1969), (1975)

$$\left\{ K \, \emptyset_x - \frac{\gamma + 1}{2} \, \emptyset_x^2 \right\}_x + \emptyset_{\bar{y}\bar{y}} + \emptyset_{\bar{z}\bar{z}} = 0, \tag{26a}$$

or, alternatively as

$$\left\{ K - (\gamma+1) \, \emptyset_x \right\} \emptyset_{xx} + \emptyset_{\bar{y}\bar{y}} + \emptyset_{\bar{z}\bar{z}} = 0. \tag{26b}$$

Here $K$ is a transonic similarity parameter defined in terms of the thickness ratio $\tau$ by

$$K = (1 - M_\infty^2) / \tau^{2/3}, \quad \bar{y} = y \tau^{1/3}, \quad \bar{z} = z \tau^{1/3},$$

$$\text{(27)}$$

and a double limit process is considered such that $\tau \longrightarrow 0$, $M_\infty \longrightarrow 1$, $K = $ finite.

The perturbation potential $\emptyset$ is defined by

$$\emptyset_x = \frac{u - u_\infty}{u_\infty} , \quad \emptyset_y = \frac{v - v_\infty}{u_\infty} , \quad \emptyset_z = \frac{w}{u_\infty} . \tag{28}$$

The choice

$$\bar{K} = (\gamma + 1) \, \tau^{2/3} \tag{29}$$

in Eq. (17) formally leads to Eq. (26a,b). It should be noted that the definitions of $K$, $\bar{y}$, $\bar{z}$ are not unique and may be multiplied by functions like $\bar{g}(M_\infty^2) = O(1)$, where $\bar{g}(1) = 1$. Further the non-linear term $\emptyset_x \emptyset_{xx}$ is an essential feature of high speed flow and is responsible for the generation of shock waves. Equation (26a) is written in divergence form and represents the conservation of mass flux

$$\text{div} (\rho \, \vec{q}) = 0, \tag{30}$$

in transonic small perturbation theory. In addition to the above equations, the shock jump conditions derived from the full Rankine-Hugoniot relation must be appended to the above equation, for the sake of completing the list of basic equations of transonic small perturbation theory. However, it turns out that shock jump conditions are contained in (26a), which is written in conservation form.

## 1.5.1. The integro-differential equation for the velocity potential

The small perturbation differential formulation presented in the previous section may be converted into an integral equation for the velocity potential, by application of Green's theorem of potential theory, which on differentiation delivers the integral equations for the velocity components. Such a presentation was given first by Gullstrand (1951) for the symmetric three dimensional case and by Heaslet and Spreiter (1957) for the unsymmetric lifting case. Under a more general set up containing a curved shock in the three-dimensional case, these equations were derived again by Kluwick and Oswatitsch (1974) under small perturbation and recently by Schmidt von Schubert (1980) for the full potential equation. For studying the transonic far field behaviour of the velocity potential, the integro-differential equation for the velocity potential in the two dimensional nonzero and zero incidence cases were given respectively by Cole (1969) and Murman and Cole (1971) and for the three dimensional lifting case by Klunker (1971). In the present section, following Klunker (1971) we derive the integro-differential equation satisfied by the velocity potential in the three dimensional lifting case of an inviscid steady transonic flow past a thin unsymmetric wing at small incidence, with subsonic free stream Mach number. The integral equations for the velocity components are derived from them by differentiation, and presented in the next section.

We recall that if $\Phi$ and $\Psi$ are functions twice continuously differentiable in a simply connected region V having S as its bounding surface, then according to Green's theorem

$$\int_V ( \Psi \nabla^2 \Phi - \Phi \nabla^2 \Psi ) \, dV = \int_S ( \Phi \frac{\partial \Psi}{\partial n} - \Psi \frac{\partial \Phi}{\partial n} ) \, dS, \qquad (31)$$

where n denotes the inward normal to S and $\nabla^2$ denotes the Laplacian operator. Identifying $\Phi$ with the reduced velocity potential, so that according to Eq. (21)

$$\nabla^2 \Phi = \frac{\partial}{\partial X} ( \frac{1}{2} \Phi^2_X ), \qquad (32)$$

and choosing $\Psi$ as the elementary solution of the Laplaces equation

$$\Psi = 1/(4\pi R), \qquad (33)$$

where

$$R = \left[ ( X - \xi)^2 + (Y - \eta)^2 + (Z - \zeta)^2 \right]^{1/2},$$

Green's theorem is applied to the region V outside the wing from which the singularities of these functions have been excluded by suitable small indentations.

The bounding surface S consists of (1) the surface at infinity chosen as a spherical surface of large radius, (2) the spherical surface $S_P$ of small radius $R_0$ around the pivotal point (X, Y, Z) which is a singularity of the function $\Psi$, (3) any shock surfaces $S_D$ on the upper and lower sides of the wing, which are surfaces of discontinuity for the normal derivatives of $\Phi$ , (4) the wing surface $S_W$ and (5) the trailing vortex sheet $S_V$ , leaving the trailing edge of the wing, assumed to lie in the XZ-plane in accordance with

the small perturbation theory. The bounding surfaces are shown in
the figure 1.2, where the directions of the respective normals are
also shown.

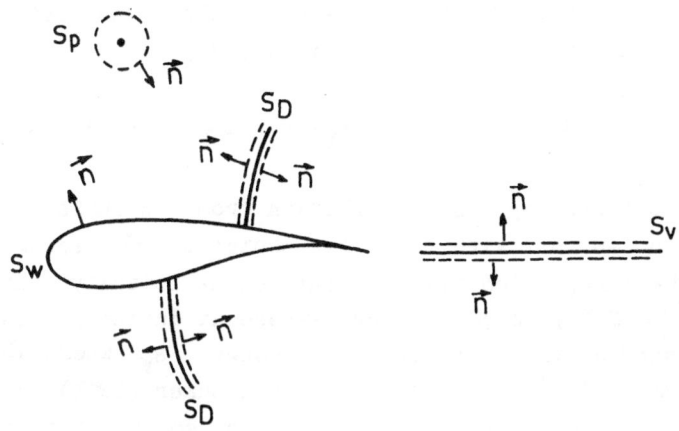

Fig.1.2    Region and surfaces of integration
          for the application of Green's theorem

The volume integral appearing on the left of Eq.(31) is simplified using equations (32) and (33).  Since $\nabla^2 \psi = 0$  in the region V,
the only contribution to the left hand side of Eq.(31) comes from the
term $\psi \nabla^2 \phi$  , which is simplified using Eq.(32) and carrying out an
integration by parts in the  x-direction.  The integration by parts
delivers integrals over the bounding surface of the volume  V , all of
which vanish except those over any shock discontinuity, denoted by  $S_D$
in the figure 1.2.   Thus,

$$\int_V \psi \nabla^2 \phi \; dV \; = \; \frac{1}{2} \int_{S_D} \psi \left[ U^2 \right] \cos \theta_1 \; dS - \frac{1}{2} \int_V U^2 \psi_\varepsilon \; dV , \qquad (34)$$

where the jump across a shock surface is

$$\left[ U^2 \right] \; = \; U_2^2 - U_1^2 , \qquad (35)$$

with subscript  1  and  2  referring to the upstream and downstream
values respectively, at the shock discontinuity surface  $S_D$ .  The

upstream unit vector normal to the shock has been taken as

$$\vec{n} = \vec{i} \cos \theta_1 + \vec{j} \cos \theta_2 + \vec{k} \cos \theta_3 , \qquad (36)$$

where ( $\vec{i}$ , $\vec{j}$ , $\vec{K}$ ) denote unit vectors along the rectangular Cartesian coordinate directions and in Eq.(34)

$$\cos \theta_1 \, dS = \text{sgn} \, (\cos \theta_1) \, d\eta \, d\zeta = - \, d\eta \, d\zeta . \qquad (37)$$

For calculating the contribution from the surface at infinity, it is to be noted that with the above choice of $\Psi$ according to Eq. (33), the integration over the surface at infinity vanishes, provided $\Phi \sim R^{-\epsilon}$ , $\epsilon > 0$ , everywhere at infinity except in a region surrounding the trailing vortex-sheet $S_V$ where $\Phi$ is required to be antisymmetric in Y . As shown by Klunker (1971), the far-field behaviour of $\Phi$ exhibits these properties, so that this integral need not be considered further.

On the spherical surface $S_P$ , $R = R_0$ ,

$$\Psi = \frac{1}{4 \pi R_0} , \qquad \frac{\partial \Psi}{\partial n} = - \frac{1}{4 \pi R_0^2} ,$$

and the element of surface $dS = R_0^2 \, d\omega$ , where $\omega$ is the solid angle, so that

$$\int_{S_P} ( \Phi \frac{\partial \Psi}{\partial n} - \Psi \frac{\partial \Phi}{\partial n} ) \, dS = \lim_{R_0 \to 0} \frac{1}{4\pi} \int_0^{4\pi} \left[ - \frac{1}{R_0^2} \Phi - \frac{1}{R_0} \frac{\partial \Phi}{\partial n} \right] R_0^2 \, d\omega$$

$$= - \Phi (X, Y, Z) . \qquad (38)$$

For calculating the integral over the shock surfaces $S_D$ , it is to be noted that $\Psi$ , $\frac{\partial \Psi}{\partial n}$ and $\Phi$ are continuous across the surfaces, whereas the normal derivative of the potential $\frac{\partial \Phi}{\partial n}$ experiences jump. Therefore

$$\int_{S_D} ( \Phi \frac{\partial \Psi}{\partial n} - \Psi \frac{\partial \Phi}{\partial n} ) \, dS = \int_{S_D} \Psi \left[ \frac{\partial \Phi}{\partial n} \right] dS , \qquad (39)$$

where $\frac{\partial}{\partial n}$ is in the direction of the upstream normal to $S_D$. It is convenient to express $\left[\dfrac{\partial \Phi}{\partial n}\right]$ as

$$\left[\frac{\partial \Phi}{\partial n}\right] = -\left[\Phi_\varepsilon\right]\cos\theta_1 - \left[\Phi_\eta\right]\cos\theta_2 - \left[\Phi_\zeta\right]\cos\theta_3.$$

(40)

Across the lifting wing surface $\Phi$ and $\dfrac{\partial \Phi}{\partial n}$ experience jump. For a thin wing $\dfrac{\partial}{\partial n} = \pm \dfrac{\partial}{\partial \eta}$. on the wing surface situated approximately along the XZ-plane. Using the notation $\Delta\Phi$ and $\Delta\dfrac{\partial \Phi}{\partial \eta}$ to denote respectively the values of $\Phi$ and $\dfrac{\partial \Phi}{\partial \eta}$ on the upper surface minus the value on the lower surface, the contribution from the integral over the wing surface follows as

$$\int_{S_W} (\Phi \frac{\partial \Psi}{\partial n} - \Psi \frac{\partial \Phi}{\partial n})\, dS = \int_{S_W} (\frac{\partial \Psi}{\partial \eta}\Delta\Phi - \Psi\Delta\frac{\partial \Phi}{\partial \eta})\, dS.$$

(41)

For a lifting surface, there exists a trailing vortex sheet $S_V$ which extends downstream from the trailing edge of the wing. In the present small disturbance approximation, we assume it to lie in the XZ-plane. The quantities $\Psi$, $\dfrac{\partial \Psi}{\partial n}$ and $\dfrac{\partial \Phi}{\partial n}$ are continuous across $S_V$ while $\Phi$ is discontinuous.

Thus

$$\int_{S_V} (\Phi \frac{\partial \Psi}{\partial n} - \Psi \frac{\partial \Phi}{\partial n})\, dS = \int_{S_V} \frac{\partial \Psi}{\partial \eta}\Delta\Phi\, dS.$$

(42)

Collecting the relations (34), (38), (39), (41) and (42), it follows from Eq. (31)

$$\Phi = \frac{1}{2}\int_V U^2 \Psi_\varepsilon\, dV + \int_{S_W} (\frac{\partial \Psi}{\partial \eta}\Delta\Phi - \Psi\Delta V)\, dS + \int_{S_V} \frac{\partial \Psi}{\partial \eta}\Delta\Phi\, dS$$

$$- \int_{S_D} \Psi \left[\{[U] - [\tfrac{1}{2}U^2]\}\cos\theta_1 + [V]\cos\theta_2 + [W]\cos\theta_3\right] dS.$$

(43)

Now, direct expansion of the shock polar relation delivers under small disturbance approximation in reduced variables (Oswatitsch 1977)

$$\left\{\left[\, U\,\right] - \left[\tfrac{1}{2}\, U^2\,\right]\right\}\left[\, U\,\right] + \left[\, v\,\right]^2 + \left[\, w\,\right]^2 = 0 \tag{44}$$

Further, the application of the divergence theorem to the irrotationality conditions, Eq. (12b), yield in reduced variables

$$\left[\, v\,\right]\cos\theta_1 - \left[\, U\,\right]\cos\theta_2 = 0, \quad \left[\, w\,\right]\cos\theta_1 - \left[\, U\,\right]\cos\theta_3 = 0,$$

$$\left[\, w\,\right]\cos\theta_2 - \left[\, v\,\right]\cos\theta_3 = 0.$$

$$\tag{45}$$

Combining relations (44) and (45) follows

$$\left\{\left[\, U\,\right] - \left[\tfrac{1}{2}\, U^2\,\right]\right\}\cos\theta_1 + \left[\, v\,\right]\cos\theta_2 + \left[\, w\,\right]\cos\theta_3 = 0,$$

$$\tag{46}$$

which is an alternative form of the shock jump conditions for weak jumps. In view of Eq.(46), the contribution from the shock surfaces to the potential $\phi$ , represented by the last integral on the right of Eq.(43) vanishes.

The terms in Eq.(43) with integrands $\frac{\partial \Psi}{\partial \eta}\, \Delta\phi$ correspond to doublets and the region of integration is over both the wing $S_W$ and the trailing vortex sheet $S_V$. These integrals may be easily transformed to one extending only over the wing surface $S_W$ through integration by parts in the $\xi$-direction and noting that $\Delta\phi$ is a function of $\zeta$ only on $S_V$ . Such reductions are common in subsonic lifting surface theory (see for example, Niyogi 1979, section 1.3), and we finally obtain

$$\phi(X, Y, Z) = \frac{Y}{4\pi} \int_{S_W} \frac{\Delta U}{Y^2 + (Z - \zeta)^2}\left[1 + \frac{X - \xi}{\left\{(X-\xi)^2 + Y^2 + (Z-\zeta)^2\right\}^{1/2}}\right] dS$$

$$- \int_{S_W} \Psi\, \Delta V\, dS + \int_V \tfrac{1}{2}\, U^2\, \Psi_\xi\, dV. \tag{47}$$

The first term on the right of Eq.(47) corresponds to a vortex distribution and represents the lifting effect, the second term represents a source distribution, whereas the third term which originates from the nonlinearity of the gasdynamic equation, has the form of a doublet with its axis in the stream direction and with a strength given by the local value of $U^2$. This term influences the results for both lifting and nonlifting wings.

The corresponding relation for the two dimensional case is obtained from Eq.(47) by regarding $\Delta U$, $U$ and $\Delta V$ on the wing to be functions of $\xi$ only and integrating with respect to $\zeta$ from $-\infty$ to $\infty$. Thus for a steady two dimensional transonic flow

$$\Phi(X, Y) = \frac{1}{2\pi} \int_c \Delta U \left\{ \frac{\pi}{2} \text{sgn}(Y) + \tan^{-1}\left(\frac{X-\xi}{Y}\right) \right\} d\xi$$

$$+ \frac{1}{2\pi} \int_c \Delta V \log \left\{ (X-\xi)^2 + Y^2 \right\}^{1/2} d\xi$$

$$+ \frac{1}{2\pi} \int_s \frac{1}{2} U^2 \frac{X-\xi}{(X-\xi)^2 + (Y-\eta)^2} dS , \qquad (48)$$

where the airfoil section is denoted by $c$, and $s$ is the two dimensional region exterior to the airfoil. The term sgn $(Y)$ in the first integral accounts for the jump in potential downstream of the trailing edge on $Y = 0$, the principal value being taken for the inverse tangent function.

Equations (47) and (48) are the basic integral relations for the velocity potential respectively for the three dimensional and two dimensional lifting cases. However, it appears convenient to work with the velocity components, rather than the velocity potential. This is achieved through differentiation, which is taken up in the section 1.6.

## 1.5.2 Velocity potential for the far field

Far away from the wing, where the distances from the wing are large compared to the dimensions of the wing, the integrals in Eqs. (47,48) may be approximated, yielding analytical expressions for the far field potential. Such representations have proved to be very useful in the finite-difference relaxation procedures in transonic aerodynamics. Following Klunker (1971), we give here the relevant representations of the far field potential, and refer to the original work for their derivation

$$\Phi \Big]_{\text{far field}} \sim \Phi_{\text{lift}} - \frac{X}{2\pi R^3} \int_{S_W} F(\mathcal{E}, \eta)\, dS + \int_{V_I} \frac{1}{2} U^2 \Psi_{\mathcal{E}}\, dV,$$

(49)

where $F(\mathcal{E}, \eta)$ denotes the wing thickness distribution. The volume integral extends over a finite interior region $V_I$, instead of an infinite region, since $U$ decays rapidly away from the wing and contributions to the volume integral from outside this region are negligibly small. The quantity $\Phi_{\text{lift}}$ denotes the contribution from the integral associated with the lift in Eq.(47). In the region where the distance from the wing $(X^2 + Y^2 + Z^2)^{1/2}$ is large with both $Y$ and $Z$ exterior to a region $O(b)$ surrounding the trailing vortex sheet, it is

$$\Phi_{\text{lift}} \sim \frac{1}{4\pi} \frac{Y}{Y^2 + Z^2}(1 + \frac{X}{R})\int_{S_W} \Delta U\, dS,$$

(49a)

whereas in the far field near the trailing vortex sheet

$$\Phi_{\text{lift}} \sim \begin{cases} \frac{1}{2}\Gamma(Z), & Y \to 0+, \quad -\frac{b}{2} \le Z \le \frac{b}{2}, \\[2ex] -\frac{1}{2}\Gamma(Z), & Y \to 0-, \quad -\frac{b}{2} \le Z \le \frac{b}{2}, \end{cases}$$

(49b)

and $\Gamma$ denotes the circulation

$$\Gamma(\eta) = \int_c \Delta U(\mathcal{E}, \eta)\, d\mathcal{E},$$

and the integration is over the chord.

The far field representation for the two dimensional case may be derived from Eq.(48) as

$$\Phi \sim \frac{\Gamma}{2\pi} \left\{ \frac{\pi}{2} \, \text{sgn} \, (Y) + \tan^{-1}(-\frac{X}{Y}) \right\} + \frac{1}{\pi} \, \frac{X}{X^2 + Y^2} \int_c F(\Xi) \, d\Xi$$

$$+ \frac{1}{2\pi} \int_{S_I} \frac{1}{2} U^2 \, \frac{X - \Xi}{(X - \Xi)^2 + (Y - \eta)^2} \, dS,$$

$$(50)$$

where  c  designates the airfoil section and  $S_I$  is the two dimensional equivalent of region  $V_I$ .

## 1.6   Integral equation formulation for three-dimensional lifting wings

The basic integral relation (47) for the reduced velocity potential of a thin lifting wing, contains the derivatives of the potential
under the integral sign and as such, it is an integro-differential
equation for the reduced velocity potential.  Although it is quite
conceivable to solve this equation by straight forward numerical means
on an electronic digital computer, no such attempts have been made so
far.  On the other hand, all the authors find it convenient to work
instead with the velocity components.  The relevant integral equations
may be derived through differentiation from (47) for the three dimensional case, taking appropriate care of the dipole singularity in the
integrand of the volume integral at the pivotal point (X, Y, Z).  A
coupled pair of nonlinear singular integral equations results, whose
form depends on the definition of the principal value of  the singular
integral.  The two-dimensional case may be derived from Eq.(48),
through differentiation.

An alternative way of arriving at the integral equation formulation is due to Nørstrud (1968, 1970), (1973a), who treated the small
perturbation gasdynamic equation (21a,b) as a Poisson equation, and
used the corresponding integral representation (Courant and Hilbert
1962) for the reduced velocity potential.  It was then differentiated
to obtain the integral equations for the velocity components.  Both
the approaches are discussed here.

Before carrying out the differentiation, it is convenient to
note a few mathematical points about singular integrals, discussed in
Mikhlin (1965) and in Ogana (1979).  Let  f  be a function which
becomes infinite at a point  P  within the volume  V.  To define the

integral of f throughout the volume V , the singular point P is
surrounded by a small enclosed cavity $\sigma$ , and allow $\sigma$ to vanish
while always surrounding P. Then the volume integral is defined as

$$\int_V f \, dV \equiv \lim_{\sigma \to 0} \int_{V-\sigma} f \, dV. \qquad (51)$$

The volume integral is said to be convergent, if the limit on the right
hand side of Eq.(51) is finite. Otherwise, it is divergent. If the
integral is convergent, but the value of the limit depends on the shape
of $\sigma$ , it is often said to be semiconvergent (Leathem 1960). Convergence of the volume integral is guaranteed if within a sphere of
finite radius with centre P , the function f satisfies the inequality

$$|f| < M R^{-\mu}, \qquad (52)$$

where $\mu < 3$ , M is a definite constant and R is the distance
between P and the point at which f is estimated. For divergence,
$\mu \geqslant 3$ , but semiconvergence may occur if $\mu = 3$.

Further, let $\psi = \psi(\mathcal{E}, \eta, \varsigma)$ be finite and have uniformly
continuous derivatives in the region V and let $\emptyset = \emptyset(\mathcal{E}-x, \eta-y, \varsigma-z)$
be infinite at the point P(x, y, z). Then the following result holds
(Leathem 1960):

$$\frac{\partial}{\partial x} \left[ \int_V \psi(\mathcal{E}, \eta, \varsigma) \, \emptyset(\mathcal{E} - x, \eta - y, \varsigma - z) \, dV \right]$$

$$= \int_V \psi(\mathcal{E}, \eta, \varsigma) \frac{\partial}{\partial x} \left[ \emptyset(\mathcal{E} - x, \eta - y, \varsigma - z) \right] dV$$

$$- \psi(x, y, z) \left\{ \lim_{\sigma \to 0} \int_\sigma \emptyset(\mathcal{E} - x, \eta - y, \varsigma - z) \, n_x \, d\sigma \right\},$$

$$(53)$$

where $dV = d\mathcal{E} \, d\eta \, d\varsigma$ and $n_x = - \cos(n, x)$, the direction of
n being that of the inward normal to the boundary. Similar result
may be established for singular surface integrals (Mikhlin 1965 ).

In his pioneering work on integral equation method, Oswatitsch (1950) defined the principal value of a singular double integral by surrounding the singularity by means of an infinite slit of small width parallel to the $\eta$-axis and taking the limit as the width of the slit approaches zero. Thus according to Oswatitsch principal value definition

$$\underset{-\infty}{\overset{\infty}{\int}}\underset{|\cdot|}{\overset{\infty}{-\infty}}\int F(\xi, \eta)\, d\xi\, d\eta = \underset{\varepsilon \to 0}{\lim} \left[ \underset{\eta=-\infty}{\overset{\infty}{\int}} \left\{ \underset{\xi=-\infty}{\overset{X-\varepsilon}{\int}} + \underset{X+\varepsilon}{\overset{\infty}{\int}} F(\xi, \eta) d\xi \right\} d\eta \right]$$

(54)

where the singularity of the integrand $F(\xi, \eta)$ is at $(X, Y)$, and the symbol $'|\cdot|'$ under the integral sign is distinctive of the Oswatitsch principal value definition. Another principal value definition is frequently used in potential theory, according to which the singularity is surrounded by a circle of small radius around it and taking the limit as the radius of the circle tends to zero. This is distinguished from the Oswatitsch principal value definition, by putting the symbol $\odot$ under the integral sign. Both of these principal value definitions have found wide use in the integral equation method and have been generalized to the three dimensional cases, the cavity corresponding to the Oswatitsch principal value definition, being two infinite planes perpendicular to the $\xi$-direction enclosing the singularity. This definition has been used in the integral equation formulations of Gullstrand (1951), Heaslet and Sprieter (1955), Nørstrud (1970), (1973a), whereas Klunker (1971), enclosed the singularity by a small sphere, as presented in the previous section.

The volume integral in Eq.(47) is semi-convergent, as shown by Ogana (1979), who calculated its partial derivative with respect to $X$, by surrounding the singularity by means of an infinitesimal hexahedron, from which the results corresponding to the above two principal value definitions follow as particular cases. Using the Oswatitsch principal value definition, it follows from Eq.(47) on differentiation with respect to $X$, the integral equation derived by Heaslet and Spreiter (1957):

$$U(X, Y, Z) = U_H(X, Y, Z) + \frac{1}{2} U^2 (X, Y, Z)$$

$$- \underset{-\infty}{\overset{\infty}{\int}}\underset{-\infty}{\overset{\infty}{\int}}\underset{|\cdot|}{\overset{\infty}{-\infty}}\int \frac{1}{2} U^2(\xi, \eta, \zeta) K_3(\xi - X, \eta - Y, \zeta - Z) d\xi\, d\eta\, d\zeta ,$$

(55a)

where $K_3$ is the three dimensional kernel having a dipole singularity at $(X, Y, Z)$

$$K_3(\mathcal{E}-X, \eta-Y, \zeta-Z) = \frac{1}{4\pi} \frac{2(\mathcal{E}-X)^2-(\eta-Y)^2-(\zeta-Z)^2}{\left[(\mathcal{E}-X)^2+(\eta-Y)^2+(\zeta-Z)^2\right]^{5/2}},$$

(55b)

and $U_H(X, Y, Z)$ is the harmonic function

$$U_H(X,Y,Z) = \frac{\partial}{\partial X}\left[\frac{Y}{4\pi}\int_{S_W}\frac{\Delta U}{Y^2+(Z-\zeta)^2}\left\{1+\frac{X-\mathcal{E}}{\sqrt{(X-\mathcal{E})^2+Y^2+(Z-\zeta)^2}}\right\}dS\right.$$

$$\left. -\frac{1}{4\pi}\int_{S_W}\frac{\Delta V}{\sqrt{(X-\mathcal{E})^2+Y^2+(Z-\zeta)^2}}dS\right].$$

(55c)

The first integral on the right of Eq.(55c) contributes to the lifting effect, whereas the second to the thickness effect. For the purely symmetrical problem of flow past a thin symmetrical wing at zero incidence, the first term vanishes identically, and $\Delta V$ is known by the tangency boundary condition, so that $U_H(X, Y, Z)$ may then be identified with the wellknown Prandtl solution (Oswatitsch 1976) which is denoted here by $U_P(X, Y, Z)$. Thus for a purely symmetrical problem, we obtain the following three dimensional non-linear singular integral equation:

$$U(X, Y, Z) = U_P(X, Y, Z) + \frac{1}{2}U^2(X, Y, Z)$$

$$-\int_{-\infty}^{\infty}\int_{-\infty}^{\infty}\int_{-\infty}^{\infty}\frac{1}{2}U^2(\mathcal{E}, \eta, \zeta)K_3(\mathcal{E}-X, \eta-Y, \zeta-Z)d\mathcal{E}\,d\eta\,d\mathcal{E},$$

(56)

where $K_3$ is given by (55b).

On the other hand, removing the singularity by a small sphere, it follows (Ogana 1979) for the lifting case

$$U(X, Y, Z) = U_H(X, Y, Z) + \frac{1}{6}U^2(X, Y, Z)$$

(57)

$$-\int_{-\infty}^{\infty}\int_{-\infty}^{\infty}\int_{-\infty}^{\infty}\frac{1}{2}U^2(\mathcal{E}, \eta, \zeta)K_3(\mathcal{E}-X, \eta-Y, \zeta-Z)d\mathcal{E}\,d\eta\,d\mathcal{E},$$

and for the corresponding symmetrical problem

$$U(X, Y, Z) = U_p(X, Y, Z) + \frac{1}{6} U^2(X, Y, Z)$$

$$- \int_{-\infty}^{\infty} \int_{-\infty}^{\infty} \int_{-\infty}^{\infty} \frac{1}{2} U^2(\xi, \eta, \zeta) K_3(\xi - X, \eta - Y, \zeta - Z) \, d\xi \, d\eta \, d\zeta ,$$

$$\tag{58}$$

where $U_H$ is defined by Eq. (55c).

It should be noted that each of the integral equations (55), (56), (57) and (58) are valid for shock free flow as well as for flows with weak shocks. It was shown by Kluwick and Oswatitsch (1974) and independently by Chakraborty (1978), that contrary to what have been stated by authors like Ferrari and Tricomi (1968), the weak shocks need not be straight and normal. However, it is not difficult to see that across straight normal shocks the volume integral in Eq.(56) is continuous, whereas across curved shocks, it was established by Kluwick and Oswatitsch (1974) by studying in details the continuity property of the shock integrals that the volume integral is no more continuous and deliver the requisite jump conditions for curved shocks. Such properties have also been studied by Schmidt von Schubert (1980) for the integral equation derived for irrotational flow past blunt bodies in three dimensions.

Due to the presence of the unknown quantity $\Delta U$ in the integrand of the first integral in Eq.(55c) defining the harmonic function $U_H(X, Y, Z)$, the integral equations for the lifting case, Eq.(55) or Eq.(57) are not readily amenable to solution. Since in aerodynamic applications, the surface pressure distribution is of greatest iterest, Nixon (1974) found it convenient to differentiate Eq.(47) with respect to $Y$ as well, take the limits as $Y \to 0+$ and $Y \to 0-$. After some manipulations, in the same way as carried out in Nixon and Hancock (1974) for the two dimensional shock-free case and presented in the section 1.7.2, Nixon established a coupled system of three-dimensional nonlinear singular integral equations for the shock-free lifting case. Since these equations have not been used for solving the three dimensional lifting problem, we do not discuss it further. It is to be noted however, that although Nixon (1974) and Nixon and Hancock (1974) have started from the hypothesis of shock-free flows the final forms of the integral equations derived by them remain valid for flows with weak shocks as well, according to our observation in the previous paragraph.

A second integral equation formulation for three-dimensional
lifting wings is due to Nørsturd (1973a), which has been used by him
for calculating transonic flow past lifting wings. A convenient way
of arriving at his formulation has been put forward by Chakraborty
(1978), which we present below. For this we note that, the harmonic
solution of the linearized subsonic theory, may be put to the form,
analogus to the non-linear representation, Eq.(47):

$$\Phi_P = \frac{Y}{4\pi} \int_{S_W} \frac{\Delta U_P}{Y^2 + (Z-\varsigma)^2} \left[ 1 + \frac{X - \varepsilon}{\sqrt{(X-\varepsilon)^2 + Y^2 + (Z-\varsigma)^2}} \right] dS$$

$$- \int_{S_W} \Psi \, \Delta V_P \, dS. \tag{59}$$

The Prandtl solution $\Phi_P$ satisfies the same tangency boundary condition as $\Phi$ corresponding to the non-linear theory, so that in accordance with thin airfoil theory

$$\Delta \Phi_Y (X, 0, Z) = \Delta \Phi_{PY} (X, 0, Z) \quad \text{at the wing planform } S_W,$$

$$\tag{60}$$

suffix Y denoting differentiation. Since the integral over $S_W$
extends over the wing planform only, the second terms on the right-
hand sides of Eq. (47) and (59) are equal to each other in view of
the tangency boundary condition, Eq. (60). Then subtracting Eq.(59)
from Eq. (47) follows

$$\Phi (X, Y, Z) = \Phi_P(X, Y, Z) + \frac{Y}{4\pi} \int_{S_W} \frac{\Delta U - \Delta U_P}{Y^2 + (Z-\varsigma)^2} \left[ 1 + \frac{X-\varepsilon}{\sqrt{(X-\varepsilon)^2 + Y^2 + (Z-\varsigma)^2}} \right] dS$$

$$+ I(X, Y, Z; U), \tag{61a}$$

where

$$I (X, Y, Z ; U) = \int_V \frac{1}{2} U^2 \Psi_\varepsilon \, dV. \tag{61b}$$

Differentiating Eq. (61a) with respect to $X$, we get

$$U(X, Y, Z) = U_P(X, Y, Z) + \frac{Y}{4\pi} \int_{S_W} \frac{\Delta U(\varepsilon, 0, \varsigma) - \Delta U_P(\varepsilon, 0, \varsigma)}{\left[ (X-\varepsilon)^2 + Y^2 + (Z-\varsigma)^2 \right]^{3/2}} \, dS + \frac{\partial I}{\partial X}.$$

$$\tag{62}$$

Now taking the limits as $Y \rightarrow \pm 0$, and considering the second term on the right-hand side of Eq.(62) as the well-known Poisson's integral, it follows

$$U(X, \pm 0, Z) = U_p(X, \pm 0, Z) \pm \frac{1}{2}\left[\Delta U(X, 0, Z) - \Delta U_p(X, 0, Z)\right]$$

$$+ \quad \lim_{Y \rightarrow \pm 0} \quad \frac{\partial I}{\partial X}(X, Y, Z; U), \qquad (63)$$

where the upper and lower signs hold respectively along the upper and lower surfaces of the wing. As already mentioned, according to Ogana (1979), the two different forms of the derivative $\frac{\partial I}{\partial X}$ are given explicitly as

$$\frac{\partial I(X, Y, Z; U)}{\partial X} = \frac{1}{2} U^2(X, Y, Z)$$

$$- \int_{-\infty}^{\infty} \int_{-\infty}^{\infty} \int_{-\infty}^{\infty} \frac{1}{2} U^2(\varepsilon, \eta, \varsigma) K_3(\varepsilon - X, \eta - Y, \varsigma - Z) \, d\varepsilon \, d\eta \, d\varsigma$$
$$|\cdot| \qquad\qquad\qquad\qquad (64a)$$

and

$$\frac{\partial I(X, Y, Z; U)}{\partial X} = \frac{1}{6} U^2(X, Y, Z)$$

$$- \int_{-\infty}^{\infty} \int_{-\infty}^{\infty} \int_{-\infty}^{\infty} \frac{1}{2} U^2(\varepsilon, \eta, \varsigma) K_3(\varepsilon - X, \eta - Y, \varsigma - Z) \, d\varepsilon \, d\eta \, d\varsigma,$$
$$\circledcirc \qquad\qquad\qquad\qquad (64b)$$

where the kernel $K_3$ has been defined in Eq.(55b). Using Eq.(64a) in Eq. (63), and taking limits, we obtain Nørstrud's formulation, in terms of Oswatitsch principal value definition:

$$U(X, \pm 0, Z) = U_p(X, \pm 0, Z) \pm \frac{1}{2}\left[\Delta U(X, 0, Z) - \Delta U_p(X, 0, Z)\right]$$

$$+ \frac{1}{2} U^2(X, \pm 0, Z) - \int_{-\infty}^{\infty} \int_{-\infty}^{\infty} \int_{-\infty}^{\infty} \frac{1}{2} U^2(\varepsilon, \eta, \varsigma) K_3(\varepsilon - X, \eta - Y,$$
$$|\cdot|$$
$$\varsigma - Z) \, d\varepsilon \, d\eta \, d\varsigma, \qquad (65)$$

valid for shock-free flow as also for flows with shocks.

If on the other hand, the singularity of the volume integral at (X, Y, Z) be removed by a vanishing sphere, we obtain instead of Eq. (65), the integral equations:

$$U(X, \pm 0, Z) = U_p(X, \pm 0, Z) \pm \frac{1}{2}\left[\Delta U(X, 0, Z) - \Delta U_p(X, 0, Z)\right]$$

$$+ \frac{1}{6} U^2(X, \pm 0, Z) - \int\limits_{-\infty}^{\infty}\int\limits_{-\infty}^{\infty}\int\limits_{-\infty}^{\infty} \frac{1}{2} U^2(\xi, \eta, \zeta) \cdot$$

$$\cdot K_3(\xi - X, \eta - Y, \zeta - Z) \, d\xi \, d\eta \, d\zeta \, . \qquad (66)$$

In equations (65) and (66), $U_p$ denotes the known linearized Prandtl solution and the upper sign refers to quantities on the upper side of the wing and the lower sign, that on the lower side. The appearance of the unknown field velocity distribution $U(X, Y, Z)$ under the integral sign is typical of all such integral formulations for the lifting case, including that of Nixon (1974) and creates some amount of inconvenience to the determination of their solution. Further it is to be noted that in view of the fact that

$$\lim_{Y \to 0+} \frac{\partial I(X, Y, Z; U)}{\partial X} = \lim_{Y \to 0-} \frac{\partial I(X, Y, Z; U)}{\partial X} ,$$

$$(67)$$

the two equations corresponding to the upper and the lower signs in Eq.(65) or (66) are not mutually independent, as may be easily verified by subtracting one from the other. This is however, not quite surprising since the velocity distributions on the upper and lower sides of the wing are not mutually indpendent . Incidentally, a second independent equation is delivered by the expression for the V-component of the velocity, obtained by differentiating Eq.(47) with respect to Y, along with the tangency boundary condition. However, we note that the tangency boundary condition has been utilized in determining the linearized solution $U_p$ and Eqs. (65) then completely determine the velocity distribution on the upper and lower profile sides. These points are discussed further in connection with the methods of solution in Chapter V.

### 1.7.1 Integral equation formulation for plane flow

For plane flow, relation (48) is the basic integrodifferential equation for the reduced velocity potential $\phi$ . As already observed in the three-dimensional case, for this case also the basic equation (48) appears to be quite amenable to solution by straight-forward numerical means on an electronic digital computer. However, no attempts have been made so far in this direction and all the authors prefer instead to work with the velocity components. The relevant formulation for the unsymmetric case of lifting plane flow past a thin unsymmetric airfoil at small incidence may be derived from Eq.(48) through differentiation.

There exists at present, three different formulations, two of which are due to Nørstrud, as presented in Nørstrud (1968, 1970), (1973b) and the other due to Nixon and Hancock (1974). An alternative derivation of the second formulation of Nørstrud (1973b) has been given by Niyogi (1978) which corrects a sign error in the earlier formulation. A controversy originated recently, from a criticism by Nixon (1975b) regarding the correctness of the integral equation formulations of Nørstrud for the lifting profile flow problem, which according to Nixon leads to non-unique results. Nørstrud (1976) on the other hand pointed out some mistakes in the Nixon critique, creating further confusion. The controversy was subsequently resolved when Chakraborty and Niyogi (1977) established that the first formulation of Nørstrud (1970) is equivalent to that of Nixon and Hancock (1974), while Niyogi (1978) showed that, apart from a sign error, the second formulation of Nørstrud (1973b) may be derived from that of Nixon and Hancock (1974). We present here all the three formulations.

### 1.7.2 Formulation of Nixon-Hancock

Differentiating Eq.(48) with respect to $X$ , it follows

$$U(X, Y) = \frac{Y}{2\pi} \int_0^1 \frac{\Delta U}{(X-\varepsilon)^2 + Y^2} \, d\varepsilon + \frac{1}{2\pi} \int_0^1 \Delta V \frac{(X-\varepsilon) \, d\varepsilon}{(X-\varepsilon)^2 + Y^2} + \frac{\partial I}{\partial X} ,$$

$$(68)$$

where the surface integral $I(X, Y ; U)$ is given by

$$I(X, Y; U) = \int\int_S \frac{1}{2} U^2 (\mathcal{E}, \eta) \frac{X - \mathcal{E}}{(X - \mathcal{E})^2 + (Y-\eta)^2} d\mathcal{E} \, d\eta ,$$

$$(69)$$

it being assumed that the airfoil chord is situated on the x-axis between 0 and 1 and S denotes the two dimensional Euclidian space $E_2$ from which the singular point $\mathcal{E} = X$, $\eta = Y$ has been removed by a suitable cavity and the slit $0 \leq \mathcal{E} \leq 1$, $\eta = \pm 0$ has been excluded. The value of the derivative $\frac{\partial I}{\partial X}$ depends on the choice of the shape of the cavity, and a little calculation shows that using Oswatitsch principal value definition, we obtain

$$\frac{\partial I(X, Y; U)}{\partial X} = \frac{U^2(X, Y)}{2} - \frac{1}{2\pi} \int_{-\infty}^{\infty} \int_{-\infty}^{\infty} \frac{1}{2} U^2(\mathcal{E}, \eta) \, K_2(\mathcal{E} - X, \eta - Y) d\mathcal{E} \, d\eta ,$$

$$(70a)$$

whereas removing the singularity by a small circle yields

$$\frac{\partial I(X, Y; U)}{\partial X} = \frac{U^2(X, Y)}{4} - \frac{1}{2\pi} \int_{-\infty}^{\infty} \int_{-\infty}^{\infty} \frac{1}{2} U^2(\mathcal{E}, \eta) \, K_2(\mathcal{E} - X, \eta - Y) d\mathcal{E} \, d\eta ,$$

$$(70b)$$

Here the kernel $K_2$ is given by

$$K_2(\mathcal{E} - X, \eta - Y) = \left[ (\mathcal{E} - X)^2 - (\eta - Y)^2 \right] \Big/ \left[ (\mathcal{E} - X)^2 + (\eta - Y)^2 \right]^2 .$$

$$(70c)$$

Limit is now taken as $Y \longrightarrow 0+$ and $Y \longrightarrow 0-$ in Eq.(68), where it is to be noted that $\Delta U$ and $\Delta V$ are functions of $\mathcal{E}$, defined by

$$\Delta U = U(\mathcal{E}, 0+) - U(\mathcal{E}, 0-), \qquad \Delta V = V(\mathcal{E}, 0+) - V(\mathcal{E}, 0-).$$

$$(71)$$

The only non-zero contribution to the first integral on the right of Eq.(68), can come from the singularity at $\mathcal{E} = X$, $Y = 0$, and we have

$$\lim_{Y \longrightarrow 0\pm} \frac{Y}{2\pi} \int_0^1 \frac{\Delta U(\mathcal{E}) \, d\mathcal{E}}{(X - \mathcal{E})^2 + Y^2} = \pm \frac{1}{2} \Delta U(X),$$

$$(72)$$

so that from Eq.(68), follows on taking the limit $Y \longrightarrow 0+$ and using Eq.(70b)

$$U(X,\ 0+) \ = \ \frac{1}{2}\ \Delta U(X) + \frac{1}{2\pi}\ \int_0^1 \frac{\Delta V(\mathcal{E})}{X - \mathcal{E}}\ d\mathcal{E} + \frac{1}{4}\ U^2(X,\ 0+)$$

$$- \lim_{Y \to 0+} \frac{1}{2\pi} \int_{-\infty}^{\infty} \int_{-\infty}^{\infty} \frac{1}{2}\ U^2(\mathcal{E},\ \eta)\ K_2(\mathcal{E}-X,\ \eta - Y)\ d\mathcal{E}\ d\eta\ ,$$

$$(73a)$$

and for the limit $Y \to 0-$

$$U(X,\ 0-) \ = \ -\frac{1}{2}\ \Delta U(X) + \frac{1}{2\pi}\ \int_0^1 \frac{\Delta V(\mathcal{E})\,d\mathcal{E}}{X - \mathcal{E}} + \frac{1}{4}\ U^2(X,\ 0-\ )$$

$$- \lim_{Y \to 0-} \frac{1}{2\pi} \int_{-\infty}^{\infty} \int_{-\infty}^{\infty} \frac{1}{2}\ U^2(\mathcal{E},\eta)\ K_2(\mathcal{E} - X, \eta - Y)\ d\mathcal{E}\ d\eta\ .$$

$$(73b)$$

Further, noting that

$$\lim_{Y \to 0-} \int_{-\infty}^{\infty} \int_{-\infty}^{\infty} U^2(\mathcal{E},\ \eta)K_2(\mathcal{E}-X,\ \eta - Y)\,d\mathcal{E}\ d\eta$$

$$= \lim_{Y \to 0+} \int_{-\infty}^{\infty} \int_{-\infty}^{\infty} U^2(\mathcal{E}, - \eta)K_2(\mathcal{E} - X,\ \eta - Y)\ d\mathcal{E}\ d\eta\ ,$$

$$(74)$$

it follows from Eqs.(73a) and (73b) through addition and simplification

$$\left[ U(X,\ 0+) + U(X,\ 0-) \right] - \frac{1}{4}\left[ U^2(X,\ 0+) + U^2(X,\ 0-\ ) \right]$$

$$= \frac{1}{\pi} \int_0^1 \frac{\Delta V(\mathcal{E})\,d\mathcal{E}}{X - \mathcal{E}} - \frac{1}{2\pi} \int_{-\infty}^{\infty} \int_{-\infty}^{\infty} \frac{1}{2}\left[ U^2(\mathcal{E},\eta) + U^2(\mathcal{E},-\eta) \right]$$

$$\cdot\ K_2(\mathcal{E}-X,\ \eta\ )\ d\mathcal{E}\ d\eta\ . \qquad (75)$$

Again, differentiating Eq.(48) with respect to $Y$ and carrying out a similar limiting procedure as above, we obtain on simplification

$$V(X, \ 0+) + V(X, \ 0-) = -\frac{1}{\pi} \int_0^1 \frac{\Delta U(\Xi) \ d\Xi}{X - \Xi}$$

$$-\frac{1}{2\pi} \int_{-\infty}^{\infty} \int_{-\infty}^{\infty} \left\{ U^2(\Xi, \eta) - U^2(\Xi, -\eta) \right\} \frac{(\Xi-X)\cdot\eta}{\left[ (\Xi - X)^2 + \eta^2 \right]^2} \ d\Xi \ d\eta$$

$$(76)$$

Equations (75) and (76) are the two basic equations derived by Nixon and Hancock (1974). Although, they formulated the problem for shock-free flow, these equations are valid for flows with weak shocks as well, since our basic equation (48) holds also for flows with weak shocks. Further, the double integrals in equations (75) and (76) still contain the field velocity distribution $U(\Xi, \eta)$ under the integral sign, which is typical for all the integral equation formulations for the lifting case. For solving these equations, some knowledge of the relationship between the field velocity distribution and that on the airfoil chord is needed. This is discussed further in the subsequent chapters, in connection with the methods of solution.

If the singularity at $\Xi = X$, $\eta = Y$ be removed by means of Oswatitsch principal value definition, then using Eq.(70a), we obtain in place of Eq.(75)

$$\left[ U(X, \ 0+) + U(X, \ 0-) \right] - \frac{1}{2} \left[ U^2(X, \ 0+) + U^2(X, \ 0-) \right]$$

$$= \frac{1}{\pi} \int_0^1 \frac{\Delta V(\Xi) d\Xi}{X - \Xi} - \frac{1}{2\pi} \int_{-\infty}^{\infty} \int_{-\infty}^{\infty} \frac{1}{2} \left[ U^2(\Xi, \eta) + U^2(\Xi, -\eta) \right]$$

$$\cdot K_2(\Xi - X, \eta) \ d\Xi \ d\eta, \tag{77}$$

which has the factor $\frac{1}{2}$ in place of $\frac{1}{4}$ in the second term on the left of Eq.(75). On the other hand Eq. (76) remains unchanged.

In the particular case of flow past a thin symmetrical airfoil at zero incidence, $\Delta U \equiv 0$ , and $V(X, \ 0+) = V(X, \ 0-)$. It follows then from Eq. (68) using Eq. (70a), the <u>integral equation of Oswatitsch</u>:

$$U(X, Y) = U_p(X, Y) + \frac{1}{2} U^2(X, Y)$$

$$- \frac{1}{2\pi} \int_{-\infty}^{\infty} \int_{-\infty}^{\infty} \frac{1}{2} U^2(\mathcal{E}, \eta) K_2(\mathcal{E} - X, \eta - Y) \, d\mathcal{E} \, d\eta, \qquad (78)$$

which appears in the pioneering work of Oswatitsch (1950). If the singularity at $\mathcal{E} = X$, $\eta = Y$ be removed by means of a small circle, using Eq. (70b), we obtain in place of Eq.(78) the alternative form of the integral equation of Oswatitsch:

$$U(X, Y) = U_p(X, Y) + \frac{1}{4} U^2(X, Y)$$

$$- \frac{1}{2\pi} \int_{-\infty}^{\infty} \int_{-\infty}^{\infty} \frac{1}{2} U^2(\mathcal{E}, \eta) K_2(\mathcal{E} - X, \eta - Y) \, d\mathcal{E} \, d\eta. \qquad (79)$$

The second term on the right of Eq.(79) differs from the corresponding term in Eq.(78) only by the factor $\frac{1}{4}$ in place of $\frac{1}{2}$. The alternative form, Eq.(79) was used by Schubert and Schleiff (1969) for studying the question of existence and uniqueness of transonic flows, discussed in Chapter VII.

In equations (78) and (79), $U_p(X, Y)$ denotes the linearized solution according to the theory of Prandtl:

$$U_p(X, Y) = \frac{1}{\pi} \int_0^1 V_0(\mathcal{E}) \frac{X - \mathcal{E}}{(X - \mathcal{E})^2 + Y^2} \, d\mathcal{E}, \qquad (80)$$

where $V_0(\mathcal{E})$ is an abreviation for $V(\mathcal{E}, 0+)$. It is a known quantity, determined by the tangency boundary condition Eq.(24d).

It is to be noted that for the symmetric case, Eq.(76) is identically satisfied and Eq.(78) or equivalently Eq. (79) is the basic equation for the integral equation method. It is a two dimensional non-linear singular integral equation for the unknown reduced velocity component $U(X, Y)$ parallel to the free-stream direction. Much work has been done over the past thirty years for developing efficient methods for solving them, which are discussed in the next two chapters.

## 1.7.3   First formulation of Nørstrud

Treating the gasdynamic equation (21b) as a Poisson's equation, its solution may be constructed in terms of the fundamental solution $\log_e (\frac{1}{R})$ of the homogeneous equation

$$\Phi_{XX} + \Phi_{YY} = 0, \tag{81}$$

as (Courant and Hilbert 1962):

$$\Phi (X, Y) = \bar{\bar{\Phi}} (X, Y) - \frac{1}{2\pi} \int\limits_{-\infty}^{\infty} \int\limits_{-\infty}^{\infty} \Phi_{\xi} \, \Phi_{\xi\xi} \, \ln(\frac{1}{R}) \, d\xi \, d\eta , \tag{82}$$

where $R = \left[ (X - \xi)^2 + (Y - \eta)^2 \right]^{1/2}$, suffix $\xi$- denoting differentiation and the harmonic function $\bar{\bar{\Phi}}(X, Y)$ is a solution of the Laplace's equation (81), to be determined along with the unknown $\Phi (X, Y)$ by means of boundary conditions. The harmonic function $\bar{\bar{\Phi}} (X, Y)$ should not be confused with the linearized Prandtl solution $\Phi_p(X, Y)$ which is a known quantity.

For simplifying the above problem, following Nørstrud (1968, 1970) the solutions are split-up into a symmetric part denoted by the superscript ' + ' and an antisymmetric part denoted by a superscript ' - ', as follows

$$\Phi(X, Y) = \Phi^+(X, Y) + \Phi^-(X, Y),$$

$$\bar{\bar{\Phi}} (X, Y) = \bar{\bar{\Phi}}^+(X, Y) + \bar{\bar{\Phi}}^-(X, Y). \tag{83}$$

The following relations hold in view of the definition of splitting:

$$\Phi^+(X, Y) = \Phi^+(X, - Y), \quad \Phi^-(X, Y) = - \Phi^-(X, - Y),$$

$$\tag{84}$$

$$\bar{\bar{\Phi}}^+(X, Y) = \bar{\bar{\Phi}}^+(X, -Y), \quad \bar{\bar{\Phi}}^-(X, Y) = - \bar{\bar{\Phi}}^-(X, - Y).$$

It appears convenient to work with the reduced velocity components instead of the reduced velocity potential. Differentiating both sides of Eq.(82) with respect to $X$ and $Y$ respectively and using Eq. (83), follows

$$U(X,Y) = \bar{U}(X,Y) + \frac{1}{2\pi} \int_{-\infty}^{\infty} \int_{-\infty}^{\infty} \left[ \Phi_{\varepsilon}^{+} \Phi_{\varepsilon\varepsilon}^{+} + \Phi_{\varepsilon}^{-} \Phi_{\varepsilon\varepsilon}^{+} + \Phi_{\varepsilon}^{+} \Phi_{\varepsilon\varepsilon}^{-} + \Phi_{\varepsilon}^{-} \Phi_{\varepsilon\varepsilon}^{-} \right]$$

$$\cdot \frac{\partial}{\partial \varepsilon} \left( \ln \frac{1}{R} \right) d\varepsilon \, d\eta , \tag{85a}$$

and

$$V(X,Y) = \bar{V}(X,Y) + \frac{1}{2\pi} \int_{-\infty}^{\infty} \int_{-\infty}^{\infty} \left[ \Phi_{\varepsilon}^{+} \Phi_{\varepsilon\varepsilon}^{+} + \Phi_{\varepsilon}^{-} \Phi_{\varepsilon\varepsilon}^{+} + \Phi_{\varepsilon}^{+} \Phi_{\varepsilon\varepsilon}^{-} + \Phi_{\varepsilon}^{-} \Phi_{\varepsilon\varepsilon}^{-} \right]$$

$$\cdot \frac{\partial}{\partial \eta} \left( \ln \frac{1}{R} \right) d\varepsilon \, d\eta . \tag{85b}$$

The solution at the profile axis is of much interest, and accordingly from Eq. (84) follows through differentiation:

$$\Phi_{X}^{+}(X, Y) = \Phi_{X}^{+}(X, -Y), \qquad \Phi_{X}^{-}(X, Y) = -\Phi_{X}^{-}(X, -Y),$$

$$\Phi_{XX}^{+}(X, Y) = \Phi_{XX}^{+}(X, -Y), \qquad \Phi_{XX}^{-}(X, Y) = -\Phi_{XX}^{-}(X, -Y),$$

$$\Phi_{Y}^{+}(X, Y) = -\Phi_{Y}^{+}(X, -Y), \qquad \Phi_{Y}^{-}(X, Y) = \Phi_{Y}^{-}(X, -Y). \tag{86}$$

Noting that integration between $-N$ and $+N$ of an integrand antisymmetric with respect to the x-axis vanishes, the symmetric and antisymmetric parts of the solution for $Y = 0$, may be separated out from Eqs. (85a,b), using Eq.(86). It follows then

$$\Phi_X^+(X,0) = \bar{\Phi}_X^+(X,0) + \frac{1}{2\pi} \int_{-\infty}^{\infty} \int_{-\infty}^{\infty} \left[ \Phi_\varepsilon^+ \Phi_{\varepsilon\varepsilon}^+ + \Phi_\varepsilon^- \Phi_{\varepsilon\varepsilon}^- \right] \left\{ \frac{\partial}{\partial \varepsilon} \left( \ln \frac{1}{R} \right) \right\}_{Y=0}$$

$$\cdot \, d\varepsilon \, d\eta \qquad (87a)$$

$$\Phi_Y^+(X,0) = \bar{\Phi}_Y^+(X,0), \qquad (87b)$$

$$\Phi_X^-(X,0) = \bar{\Phi}_X^-(X,0), \qquad (88a)$$

$$\Phi_Y^-(X,0) = \bar{\Phi}_Y^-(X,0)$$

$$+ \frac{1}{2\pi} \int_{-\infty}^{\infty} \int_{-\infty}^{\infty} \left[ \Phi_\varepsilon^+ \Phi_{\varepsilon\varepsilon}^- + \Phi_\varepsilon^- \Phi_{\varepsilon\varepsilon}^+ \right] \left\{ \frac{\partial}{\partial \eta} \left( \ln \frac{1}{R} \right) \right\}_{Y=0} d\varepsilon \, d\eta ,$$

$$(88b)$$

which are the basic equations derived by Nørstrud. Performing an integration by parts and using Oswatitsch principal value definition, Eq.(87a) may be put to the following form in terms of the reduced velocity components

$$U^+(X,0) = \bar{U}^+(X,0) + \frac{1}{2} \left\{ U^+(X,0) \right\}^2 + \bar{U}_A(X,0)$$

$$- \frac{1}{2\pi} \int_{-\infty}^{\infty} \int_{-\infty}^{\infty} \frac{1}{2} \left\{ U^+(\varepsilon,\eta) \right\}^2 K_2(\varepsilon-X,\eta) \, d\varepsilon \, d\eta ,$$
$$\scriptstyle |\cdot|$$

$$(89a)$$

where $\bar{U}_A(X,0)$ correspond to the antisymmetric part

$$\bar{U}_A(X,0) = \frac{1}{2} \left\{ \bar{U}^-(X,0) \right\}^2 - \frac{1}{2\pi} \int_{-\infty}^{\infty} \int_{-\infty}^{\infty} \frac{1}{2} \left\{ \bar{U}^-(\varepsilon,\eta) \right\}^2 K_2(\varepsilon-X,\eta) \, d\varepsilon \, d\eta .$$
$$\scriptstyle |\cdot|$$

$$(89b)$$

If the singularity at $\varepsilon = X$, $\eta = Y$ be removed by a small circle, then the second term on the right of Eq.(89a) and the first term on the right of Eq.(89b) will have the factor $\frac{1}{4}$ instead of $\frac{1}{2}$.

The solution $\overline{\Phi}(X, Y)$ of the linearized Eq.(81) is to be determined by means of the tangency boundary condition at the profile Eq. (24b). Differentiating Eq.(82) with respect to Y and taking the limit as $Y \rightarrow \pm 0$, follows for the upper and lower profile sides

$$V(X, \pm 0) = \overline{V}(X, \pm 0) - \frac{1}{2\pi} \int_{-\infty}^{\infty} \int_{-\infty}^{\infty} U(\mathcal{E}, \eta) \frac{\partial U(\mathcal{E}, \eta)}{\partial \mathcal{E}} \frac{\eta}{(X - \mathcal{E})^2 + \eta^2} d\mathcal{E} \, d\eta .$$

$$(89c)$$

For a purely symmetrical problem, the downwash velocity V and its linear counterpart $\overline{V}$ satisfy the same tangency boundary condition. As against this, from Eq.(89c) we see that for the lifting problem the downwash velocity V is composed of its linear counterpart $\overline{V}$ and an additional term generated by the unsymmetrical non-linear compressibility sources $U \frac{\partial U}{\partial \mathcal{E}}$ over whole space, which however, is not known a'priori and a deferred correction iterative approach is indicated for its determination.

## 1.7.4 Equivalence of the formulations of Nixon-Hancock and Nørstrud

The formulations of Nixon-Hancock and Nørstrud, presented respectively in sections 1.7.3 and 1.7.2 are apparently different. Since they describe the same problem, it is natural to expect them to be equivalent. Such a equivalence proof has been given by Chakraborty and Niyogi (1977), which is presented in this section.

We start from the Nørstrud equations (87a, b) and (88a, b). Integrating Eq.(87a) by parts with respect to $\mathcal{E}$ and excluding the singularity at $\mathcal{E} = X$, $\eta = Y$, by means of a small circle around it, yields

$$U^+(X, 0) = \overline{U}^+(X, 0) + \frac{1}{4} \left[ \left\{ U^+(X, 0) \right\}^2 + \left\{ U^-(X, 0) \right\}^2 \right]$$

$$- \frac{1}{2\pi} \int_{-\infty}^{\infty} \int_{-\infty}^{\infty} \frac{1}{2} \left[ \left\{ U^+(\mathcal{E}, \eta) \right\}^2 + \left\{ U^-(\mathcal{E}, \eta) \right\}^2 \right]$$

$$\cdot \Psi_{\mathcal{E}X} \, d\mathcal{E} \, d\eta , \qquad (90)$$

where the function $\Psi(X, \mathcal{E}; Y, \eta)$ is given by

$$\Psi(X, \mathcal{E} \, ; \, Y \, , \, \eta \,) \; = \; \ln\left[\, (X - \mathcal{E})^2 + (Y - \eta)^2 \,\right]^{1/2}, \qquad (91)$$

and suffixes $\mathcal{E}$ and $X$ denote differentiation. Now, by definition

$$U(X, \, 0+ \,) \; = \; U^+ + U^- \, , \qquad U(X, \, 0-) = \; U^+ - U^- \, , \qquad (92a)$$

so that

$$U(X, \, 0+ \,) + \; U(X, \, 0-) \; = \; 2U^+ \, ,$$

and

$$U^2(X, 0+) + U^2(X, 0-) \; = \; 2\left[\,\{U^+(X,0)\}^2 + \{U^-(X,0)\}^2\,\right] \qquad (92b)$$

$$U^2(\mathcal{E}, \, \eta) - U^2(\mathcal{E}, -\eta) \; = \; 4 \, U^+(\mathcal{E}, \, \eta) \, U^-(\mathcal{E}, \, \eta \,). \qquad (92c)$$

It consequently follows from Eq.(90), using Eq.(88a) that

$$U(X, 0+) \; = \; \bar{U}^+(X,0) + \bar{U}^-(X,0) + \frac{1}{4}\left[\,\{U^+(X,0)\}^2 + \{U^-(X,0)\}^2\,\right]$$

$$- \; \frac{1}{2\pi} \int\limits_{-\infty}^{\infty} \int\limits_{-\infty}^{\infty} \frac{1}{2}\left[\,\{U^+(\mathcal{E}, \, \eta)\}^2 + \{U^-(\mathcal{E}, \, \eta)\}^2\,\right] \Psi_{\mathcal{E}X} \; d\mathcal{E} \; d\eta,$$

$$(93a)$$

and

$$U(X, 0-) \; = \; U^+(X, \, 0) - U^-(X, \, 0)$$

$$= \; \bar{U}^+(X,0) - \bar{U}^-(X,0) + \frac{1}{4}\left[\,\{U^+(X,0)\}^2 + \{U^-(X,0)\}^2\,\right]$$

$$- \; \frac{1}{2\pi} \int\limits_{-\infty}^{\infty} \int\limits_{-\infty}^{\infty} \frac{1}{2}\left[\,\{U^+(\mathcal{E}, \eta \,)\}^2 + \{U^-(\mathcal{E}, \eta \,)\}^2\,\right] \Psi_{\mathcal{E}X} \; d\mathcal{E} \; d\eta.$$

$$(93b)$$

Adding Eqs(93) and using Eq.(92b) it follows that

$$U(X, 0+) + U(X, 0-) = \varepsilon \bar{U}^+(X, 0) + \frac{1}{4}\left[ U^2(X, 0+) + U^2(X, 0-) \right]$$

$$- \frac{1}{4\pi} \int_{-\infty}^{\infty} \int_{-\infty}^{\infty} \left[ U^2(\varepsilon, \eta) + U^2(\varepsilon, -\eta) \right] \Psi_{\varepsilon X} \, d\varepsilon \, d\eta,$$

$$\text{(94)}$$

which is identical with the first equation of Nixon-Hancock, viz. Eq.(75), where the line integral is equal to the quantity $2 \bar{U}^+(X, 0)$, that is

$$\bar{U}^+(X, 0) = \frac{1}{2\pi} \int_{0}^{1} \frac{\Delta V(\varepsilon)}{X - \varepsilon} \, d\varepsilon .$$

$$\text{(95)}$$

On the other hand, integrating Eq.(88b) by parts with respect to $\varepsilon$ yields

$$\bar{V}^-(X, 0) = \bar{V}^-(X, 0) - \frac{1}{2\pi} \int_{-\infty}^{\infty} \int_{-\infty}^{\infty} U^+(\varepsilon, \eta) \, U^-(\varepsilon, \eta) \, \Psi_{\varepsilon \eta} \, d\varepsilon \, d\eta .$$

$$\text{(96)}$$

By addition and subtraction it follows from Eqs.(96) and (88a) that

$$\bar{V}^-(X, 0) + V^+(X, 0) = \bar{V}^+(X, 0) + \bar{V}^-(X, 0)$$

$$- \frac{1}{2\pi} \int_{-\infty}^{\infty} \int_{-\infty}^{\infty} U^+(\varepsilon, \eta) \, U^-(\varepsilon, \eta) \, \Psi_{\varepsilon \eta} \, d\varepsilon \, d\eta ,$$

$$\text{(97a)}$$

and

$$\bar{V}^-(X, 0) - V^+(X, 0) = - \bar{V}^+(X, 0) + \bar{V}^-(X, 0)$$

$$- \frac{1}{2\pi} \int_{-\infty}^{\infty} \int_{-\infty}^{\infty} U^+(\varepsilon, \eta) \, U^-(\varepsilon, \eta) \, \Psi_{\varepsilon \eta} \, d\varepsilon \, d\eta .$$

$$\text{(97b)}$$

Noting that

$$V(X, 0+) = V^+ + V^- , \qquad V(X, 0-) = -V^+ + V^- ,$$

$$\text{(98)}$$

it follows from Eqs.(97) by addition, and using Eq.(92c) that

$$V(X,0+) + V(X,0-) = 2\bar{V}^{-}(X,0) - \frac{1}{4\pi} \int_{-\infty}^{\infty} \int_{-\infty}^{\infty} \left[ U^2(\varepsilon,\eta) - U^2(\varepsilon,-\eta) \right] \Psi_{\varepsilon\eta} \, d\varepsilon \, d\eta,$$

(99)

which is the second Nixon-Hancock equation (76), where the line integral has been replaced by the unknown quantity $2\bar{V}^{-}(X,0)$ ; that is

$$\bar{V}^{-}(X,0) = -\frac{1}{2\pi} \int_{0}^{1} \frac{U(\varepsilon,0+) - U(\varepsilon,0-)}{X - \varepsilon} \, d\varepsilon .$$

(100)

The steps taken in the above proof may be retraced to obtain the system (87), (88) from that of system (75), (76), showing their equivalence.

### 1.7.5 Second formulation of Nørstrud

The integral equation formulation for transonic lifting airfoils has been brought to a mathematically neat form by Nørstrud (1973b) in his second formulation. Following Niyogi (1978) we present here an alternative derivation of it.

Denoting by $U_p(X,Y)$ and $V_p(X,Y)$ the reduced velocity components corresponding to the linearized Prandtl solution, we note that $U_p(X,Y)$ satisfies the integral equation

$$U_p(X,Y) = \frac{Y}{2\pi} \int_{0}^{1} \frac{\Delta U_p(\varepsilon)}{(X-\varepsilon)^2+Y^2} \, d\varepsilon + \frac{1}{2\pi} \int_{0}^{1} \Delta V_p(\varepsilon) \frac{(X-\varepsilon)d\varepsilon}{(X-\varepsilon)^2 + Y^2} .$$

(101)

Further, noting that according to the thin airfoil theory boundary condition

$$\Delta V(X) = \Delta V_p(X), \quad 0 \leqslant X \leqslant 1 ,$$

(102)

it follows from Eq.(68) using Eqs.(101) and (102)

$$U(X,Y) = U_p(X,Y) + \frac{1}{2\pi} \int_0^1 \left[ \Delta U(\mathcal{E}) - \Delta U_p(\mathcal{E}) \right] \frac{Y\, d\mathcal{E}}{(X - \mathcal{E})^2 + Y^2} + \frac{\partial I}{\partial X} \; .$$

(103)

Observing that

$$\lim_{Y \to \pm 0} \frac{1}{2\pi} \int_0^1 \left[ \Delta U(\mathcal{E}) - \Delta U_p(\mathcal{E}) \right] \frac{Y\, d\mathcal{E}}{(X - \mathcal{E})^2 + Y^2}$$

$$= \pm \frac{1}{2} \left[ \Delta U(X) - \Delta U_p(X) \right] , \qquad (104)$$

it follows from Eq.(103), taking limits respectively as $Y \to 0+$ and as $Y \to 0-$ ,

$$U(X,0+) = U_p(X,0+) + \frac{1}{2}\left[ \Delta U(X) - \Delta U_p(X) \right] + \lim_{Y \to 0+} \frac{\partial I}{\partial X} \; ,$$

(105a)

and

$$U(X,0-) = U_p(X,0-) - \frac{1}{2}\left[ \Delta U(X) - \Delta U_p(X) \right] + \lim_{Y \to 0-} \frac{\partial I}{\partial X} \; .$$

(105b)

Using the value of the derivative $\frac{\partial I}{\partial X}$ given by Eq.(70a) corresponding to the Oswatitsch principal value definition, Eqs.(105) may be rewritten for the upper profile side as

$$U(X,0+) = U_p(X,0+) + \frac{1}{2}\left[ \Delta U(X) - \Delta U_p(X) \right]$$

$$+ \frac{1}{2}U^2(X,0+) - \frac{1}{2\pi} \int_{-\infty}^{\infty}\int_{-\infty}^{\infty} \frac{1}{2}U^2(\mathcal{E},\eta)\, K_2(\mathcal{E}-X, \eta)\, d\mathcal{E}\, d\eta ,$$

(106a)

and for the lower profile side

$$U(X, 0-) = U_p(X, 0-) - \frac{1}{2} \left[ \Delta U(X) - \Delta U_p(X) \right]$$

$$+ \frac{1}{2} U^2(X, 0-) - \frac{1}{2\pi} \int\limits_{-\infty}^{\infty} \int\limits_{-\infty}^{\infty} \frac{1}{2} U^2(\mathcal{E}, \eta) K_2(\mathcal{E}-X, \eta) \ d\mathcal{E} \ d\eta ,$$

$$(106b)$$

which are the basic equations of the second formulation of Nørstrud. There is a sign error in Nørstrud (1973b) which has been corrected here.

It may be noted that the two Eqs.(106) are not mutually independent, as may verified by subtracting one from the other, when it leads to an equation of the form 'zero equal to zero'. This is not surprising in view of the fact that the velocity distributions on the upper and lower sides of a lifting profile are not mutually independent. Incidentally, a second independent equation is delivered by the velocity component V together with the tangency boundary condition at the profile which has been incorporated in the above equations through the Prandtl solution $U_p(X, Y)$. Thus, Eqs. (106) form a self-contained system and describes the lifting problem completely. However, it suffers from the typical draw-back, present in the other formulations as well, that the double integral contains the unknown field velocity distribution $U(\mathcal{E}, \eta)$ under the integral sign. This is discussed further in Chapter IV in connection with the methods of solution.

## Chapter II

## Methods of Solution : Earlier Developments

### 2.1  Introduction

The problem formulations along with the basic equations of the
integral equation method have been presented in the previous chapter.
The methods of solution in the different cases are dealt with in  the
present and subsequent chapters.  Historically looking, the integral
equation method experienced two distinct phases of developments in
the two decades 1950 to 1962 and in 1968 to the present day.  The
pioneering work of Oswatitsch appeared in 1950, which was extended
and modified in the first phase by Gullstrand, Spreiter-Alksne,
Oswatitsch-Keune and Zierep.  The second phase started with the doc-
toral dissertation of Nørstrud, which was marked by the intensive use
of electronic digital computers.  The works  of Nixon-Hancock, Nørstrud,
Frohn, Hansen, the present author and his associates appeared in the
second phase.  The developments of the first phase, mainly methods of
pre-computer days, clearly showed the integral equation method to be
a dependable one which could deliver relatively quickly results of
moderate accuracy, whereas the developments in the second phase established it as a strong competetor of the finite-difference relaxation
procedures that deliver results of high accuracy.

In the present chapter we discuss the earlier developments.
Since the developments of this phase have already appeared in a number
of works, like Ferrari and Tricomi (1968), Zierep (1966) and Niyogi
(1977), we restrict ourselves here only to the essential features and
results obtained in this phase.  The later developments are discussed
in some details in the subsequent chapters.

### 2.2  Method of Oswatitsch for a symmetric profile at zero incidence

We consider first the simplest case of steady plane inviscid
transonic flow past a thin symmetric profile at zero incidence, with
free-stream Mach number  $M_{\infty} < 1$.  We choose a body fixed Cartesian

coordinate system as shown in Fig.2.1. Due to symmetry of the problem, it is sufficient to consider only the upper half of the profile. A typical transonic flow is considered here, so that a supersonic region embedded in an otherwise subsonic flow is formed near the maximum thickness of the profile. The supersonic region is often found to terminate in a shock, as shown in Fig.2.1(b), whereas in other cases continuous, i.e. shock-free supersonic regions may be formed as shown in Fig.2.1(a). The sonic line, which demarcates the subsonic and super-sonic regions is not known a'priori, and is to be found out as a part of the solution.

Then as shown in section 1.7.1, the basic small perturbation equations may be converted to the integral equation of Oswatitsch, expressed in reduced coordinates with Oswatitsch principal value

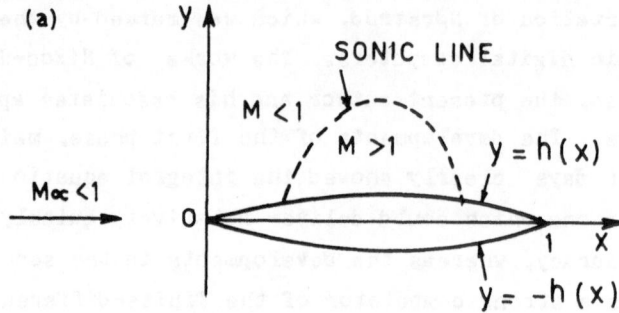

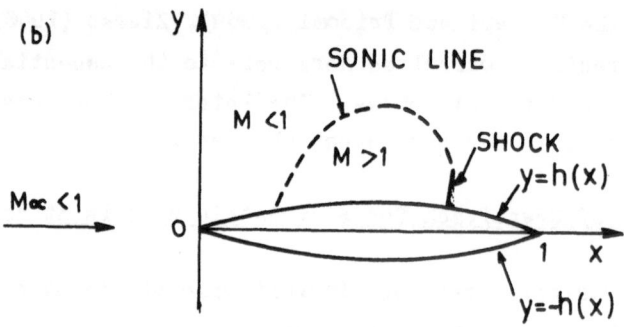

Fig.2.1  Transonic flow past a thin symmetric profile
(a) Shock-free flow, (b) flow with shock.

definition as

$$U(X, Y) = U_p(X, Y) + \frac{1}{2} U^2(X, Y)$$

$$- \frac{1}{2\pi} \int_{-\infty}^{\infty} \int_{-\infty}^{\infty} \frac{1}{2} U^2(\mathcal{E}, \eta) \frac{(\mathcal{E}-X)^2 - (\eta - Y)^2}{\left[(\mathcal{E}-X)^2 + (\eta - Y)^2\right]^2} d\mathcal{E} d\eta$$

$$|\cdot|$$

$$(1)$$

where $U_p(X, Y)$ denotes the known linearized solution due to Prandtl defined by Eq. (1.80), and repeated here for ready reference:

$$U_p(X, Y) = \frac{1}{\pi} \int_0^1 V_0(\mathcal{E}) \frac{X - \mathcal{E}}{(X - \mathcal{E})^2 + Y^2} d\mathcal{E} .$$
$$(2)$$

Here $V_0(\mathcal{E}) \equiv V(\mathcal{E}, 0)$ is a known quantity determined by the tangency boundary condition at the thin profile Eq.(1.24d)

$$V(X, 0) = T \frac{dq}{dX} , \qquad \text{on the upper part,} \quad Y = 0+ \qquad (3)$$

where $T$ denotes the reduced thickness ratio, which is a transonic similarity parameter, expressed in terms of its true value $\tau$ by

$$T = \tau \Big/ \Big[ ( \frac{1}{M_\infty^*} - 1)(1 - M_\infty^2)^{1/2} \Big] . \qquad (4a)$$

The reduced thickness ratio $T$ is a similarity parameter closely related to the transonic similarity parameter $\chi$ introduced by Spreiter (1954). In fact,

$$\chi = |1 - M_\infty^2| \Big/ \Big[ \tau(\gamma + 1)M_\infty^2 \Big]^{2/3} \approx \frac{1}{T^{2/3}} \qquad (4b)$$

The thin symmetric profile shape $g(x)$ has been taken here as

$$g(x) = \tau q(x), \quad 0 \leqslant x \leqslant 1,$$
$$= 0, \quad \text{otherwise} . \qquad (5)$$

If instead of Oswatitsch principal value definition, the singularity at $\mathcal{E} = X$, $\eta = Y$ be excluded by a small circle and the limit is taken as the radius of the circle approaches zero, the alternative form of the integral equation of Oswatitsch Eq.(1.79) is obtained in place of Eq. (1):

$$U(X, Y) = U_p(X, Y) + \frac{1}{4} U^2(X, Y)$$

$$- \frac{1}{2\pi} \int\limits_{-\infty}^{\infty} \int\limits_{-\infty}^{\infty} \frac{1}{2} U^2(\varepsilon, \eta) \frac{(\varepsilon - X)^2 - (\eta - Y)^2}{\left[ (\varepsilon - X)^2 + (\eta - Y)^2 \right]^2} d\varepsilon d\eta ,$$

$$(6)$$

which differs from Eq.(1) only in the factor of the second terms on the right. The double integrals in Eqs. (1) and (6) possess a dipole singularity at the point $\varepsilon = X, \eta = Y$ .

Our problem is to solve the two-dimensional non-linear singular integral equation (1) (or Eq.(6) ) for the unknown reduced velocity distribution $U(X, Y)$. The reduced surface pressure distribution $C_p \Big|_{\text{surface}}$ , expressed directly in terms of $U(X, 0)$ (c.f. Eq.(1.20d) ) as

$$C_p \Big]_{\text{surface}} = \frac{\bar{K}}{1 - M_\infty^2} (c_p)_{\text{surface}} = -2U(X, 0), \qquad (7)$$

is of particular interest. Putting $Y = 0$ in Eq.(1), it follows

$$U(X, 0) = U_p(X, 0) + \frac{1}{2} U^2(X, 0)$$

$$- \frac{1}{2\pi} \int\limits_{-\infty}^{\infty} \int\limits_{-\infty}^{\infty} \frac{1}{2} U^2(\varepsilon, \eta) \frac{(\varepsilon - X)^2 - \eta^2}{\left[ (\varepsilon - X)^2 + \eta^2 \right]^2} d\varepsilon d\eta , \qquad (8)$$

which however, still contains the unknown field velocity distribution $U(\varepsilon, \eta)$ under the integral sign. For solving Eq.(8), Oswatitsch (1950) first approximated it by a one-dimensional non-linear singular integral equation, now known as the simplified Oswatitsch equation.

Oswatitsch introduced in Eq.(8), the following substitution for the velocity field in terms of that on the profile axis:

$$U(X, Y) = U_0(X) \Big/ \left[ 1 + \frac{|Y|}{b(X)} \right]^2 , \qquad (9)$$

where $U_0(X) \equiv U(X, 0)$. The parameter $b(X)$ may be determined by means of the irrotationality condition

$$\frac{\partial U}{\partial Y} - \frac{\partial V}{\partial X} = 0, \qquad (10)$$

and the tangency boundary condition at the profile Eq.(3) as

$$b(X) \;=\; -2U_0(X) \Big/ \left[ T \frac{d^2 q}{dX^2} \right] \tag{11a}$$

Thus, the parameter $b$ depends also on the unknown solution $U_0(X)$, which we sometimes stress by writing

$$b(X) \;=\; b(X \,;\, U_0(X)\,). \tag{11b}$$

The substitution Eq.(9) has the desirable property that it gives the correct value of the velocity on the profile axis and decays like a dipole for large values of $Y$. Similar substitutions with various modifications have been used later by many authors, like Gullstrand (1951), Nørstrud (1970), (1973a,b), Nixon and Hancock (1974), Mitra (1976), Hansen (1976a) and critically studied by Chakraborty (1974). Generally, satisfactory results have been obtained by their use. A drawback of the above substitution immediately noticed from Eq.(11a) is that the velocity perturbations in the semi-infinite region before and after the profile vanishes identically. At lower transonic conditions where the shock is situated fairly ahead of the trailing edge, these contributions are small due to rapid attenuation of perturbations. However, with increasing Mach number, it seems that the approximation becomes worse.

The solution at the profile surface is of primary interest from the standpoint of application and it is important to represent it correctly near the profile surface than in front or aft of it, where in any case $U(X,\ Y)$ is small. Thus the variation in front or aft of the profile may be neglected. The kernel of the double integral in Eqs.(1) or (6) decays very rapidly away from the profile and any error in the representation of $U(X,\ Y)$ in the field outside the profile may be expected to be neutralized by it. Moreover, the double integral is not sensitive to small changes in $U(X,\ Y)$, particularly so, far away from the profile.

Substituting Eq.(9) in the double integral in Eq.(8) the integration with respect to $\eta$ may be carried out as follows:

$$\int\limits_{-\infty}^{\infty}\int\limits_{-\infty}^{\infty} \frac{U^2(\xi,\eta)}{2} \; \frac{(\xi-X)^2 - \eta^2}{\left[(\xi-X)^2 + \eta^2\right]^2} \, d\xi d\eta \;=\; \int\limits_{-\infty}^{\infty} U_0^2(\xi)\, d\xi \int\limits_{0}^{\infty} \frac{1}{\left(1+\frac{\eta}{b}\right)^4} \; \frac{(\xi-X)^2 - \eta^2}{\left[(\xi-X)^2 + \eta^2\right]^2} \, d\eta .$$

Putting $\eta = b\eta'$ and $(\mathcal{E} - X)/b(\mathcal{E}) = Z$ in the second integral above, performing an integration by parts with respect to $\eta$ yields

$$\int_0^\infty \frac{1}{(1 + \frac{\eta}{b})^4} \frac{(\mathcal{E}-X)^2 - \eta^2}{\left[(\mathcal{E}-X)^2 + \eta^2\right]^2} d\eta = \frac{4}{b} \int_0^\infty \frac{\eta'}{Z^2 + \eta'^2} \frac{d\eta'}{(\eta' + 1)^5} .$$

Putting again $\bar{\eta} = 1/(\eta' + 1)$, and resolving into partial fractions, the integration may be carried out. We obtain after a little calculation

$$\frac{1}{2\pi} \int_{-\infty}^\infty \int_{-\infty}^\infty \frac{1}{2} U^2(\mathcal{E}, \eta) \frac{(\mathcal{E}-X)^2 - \eta^2}{\left[(\mathcal{E}-X)^2 + \eta^2\right]^2} d\mathcal{E} d\eta$$

$$= \int_{-\infty}^\infty \frac{U_0^2(\mathcal{E})}{2b(\mathcal{E}; U_0(\mathcal{E}))} E(Z) d\mathcal{E} , \qquad (12a)$$

where

$$E(Z) = \frac{4}{\pi(1 + Z^2)^5}\left[\frac{\pi}{2}(5 - 10Z^2 + Z^4)|Z| - (1-10Z^2+5Z^4)\log_e|Z|\right.$$

$$\left. - \frac{1}{12}(25 - 71Z^2 - Z^4 - Z^6)(1 + Z^2)\right]. $$

$$(12b)$$

Thus, with the help of the substitution (9), Eq.(8) changes over to

$$U_0(X) = U_{P_0}(X) + \frac{1}{2} U_0^2(X) - \int_{-\infty}^\infty \frac{U_0^2(\mathcal{E})}{2b(\mathcal{E}; U_0(\mathcal{E}))} E\left[\frac{\mathcal{E} - X}{b(\mathcal{E}; U_0(\mathcal{E}))}\right] d\mathcal{E} .$$

$$(13)$$

It is to be noted that from Eq.(11a) follows that before and after the profile the value of the parameter $(1/b)$ is zero. Consequently the integral in Eq.(13) vanishes before and after the profile. Hence the integral equation of Oswatitsch reduces to

$$U_0(X) = U_{P_0}(X) + \frac{1}{2} U_0^2(X) - \int_0^1 \frac{U_0^2(\mathcal{E})}{2b(\mathcal{E}; U_0(\mathcal{E}))} E\left[\frac{\mathcal{E} - X}{b(\mathcal{E}; U_0(\mathcal{E}))}\right] d\mathcal{E},$$

$$(14)$$

known as the simplified Oswatitsch equation.

The kernel $E(Z)$ of the simplified Oswatitsch equation (13) possesses a logarithmic singularity at $Z = 0$ and decays rapidly with increasing $Z$, as shown in Fig.2.2.

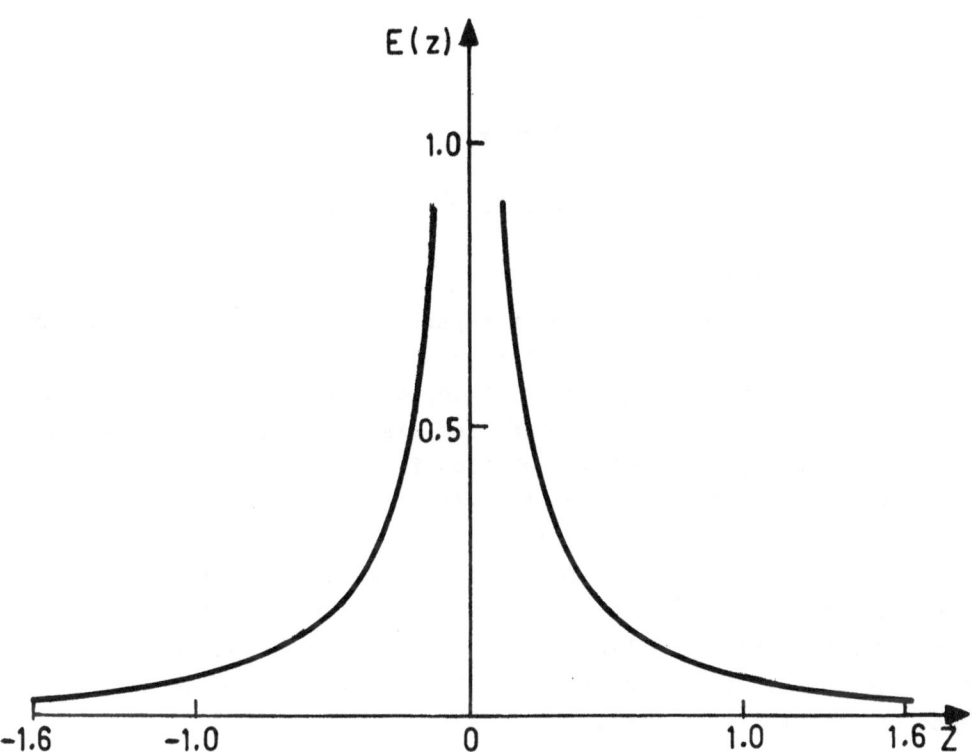

Fig.2.2  The influence function $E(Z)$.

Oswatitsch solved Eq.(14) approximately by substituting assumed forms of the solution containing a number of disposable parameters and requiring that the integral equation be satisfied at a number of important points on the body, as reported in details in Niyogi (1977). The resulting solution computed for parabolic arc profiles,

$$
g(x) = \begin{cases} 2T\ (x - x^2),\ 0 \le x \le 1, \\ \\ 0,\quad \text{otherwise}, \end{cases} \tag{15}
$$

and for NACA 0012 profiles,

$$
g(x) = \begin{cases} \dfrac{T}{0.20}(0.29620\sqrt{x} - 0.12600x - 0.35160\ x^2 \\ \quad + 0.28430\ x^3 - 0.10150\ x^4),\ 0 \le x \le 1, \\ 0,\ \text{otherwise}, \end{cases} \tag{16}
$$

showed all the typical features of supercritical transonic flow. Upto certain supercritical values of the reduced thickness ratio T , this method provides with ambiguous solution, one of which is shock-free and the other with shock. However, for higher values of the reduced thickness ratio, no shock-free solution is possible. For example, in the case of a parabolic arc profile with reduced thickness ratio T = 0.785, it predicts the possibility of an expansion shock, along with a shock-free solution. The expansion shock is contrary to the second law of thermodynamics. Whether the solution with the expansion shock exists mathematically, remained an open question. Later work on this question by Steger and Lomax (1972) shows the mathematical existence of both the solutions.

For a parabolic arc profile, Oswatitsch found that the flow becomes supercritical already at T = 0.6 and up to values of T = 0.785 supercritical shock-free solution was obtained, while for T = 0.942 no shock-free solution could be found. Solutions computed by Oswatitsch for a parabolic arc profile of reduced thickness ratios T = 0.785 and T = 0.942 are presented in Fig.2.3 and Fig.2.4, where results of other methods are also presented, for the sake of comparison. Keeping in mind that the solutions of Oswatitsch were computed in the pre-computer days, it is to be admitted that the above results of Oswatitsch are rather good. Comparison with the computations of Mitra (1976), indicate that the shock position is in fairly good agreement and the re-expansion after the shock is also well reproduced.

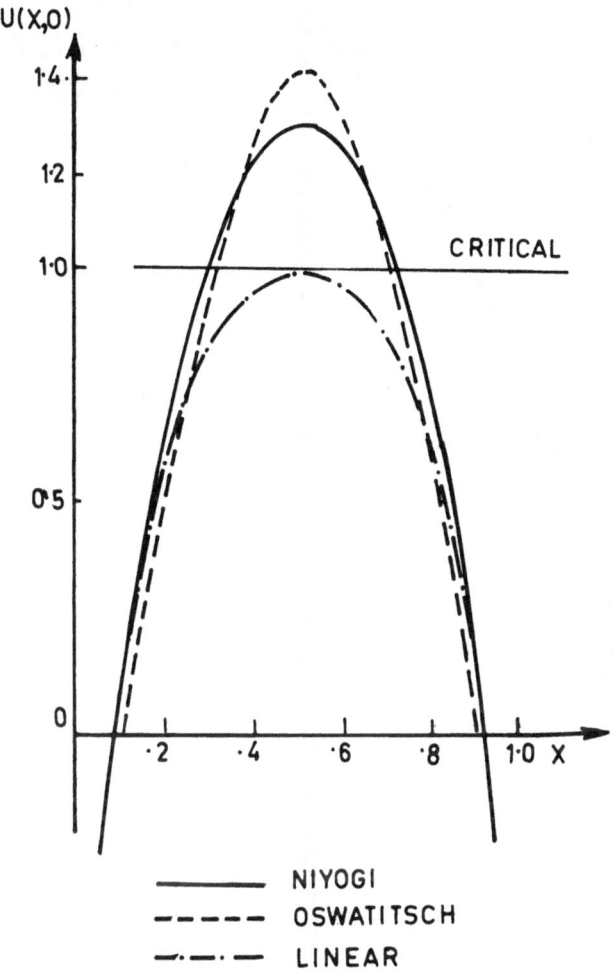

PARABOLIC ARC PROFILE,
T = 0·785

——————— NIYOGI
– – – – – OSWATITSCH
–·—·— LINEAR

Fig. 2.3    Reduced velocity distribution of a
parabolic arc profile with T = 0.785
as computed by Oswatitsch and comparison
with other results.

## 2.3    Iterative computation of Spreiter-Alksne

Spreiter and Alksne (1955) on the other hand solved the simplified Oswatitsch equation (14) iteratively. Their procedure has been
reported in full details in Ferrari and Tricomi (1968) and we present
here only the main aspects of the procedure.

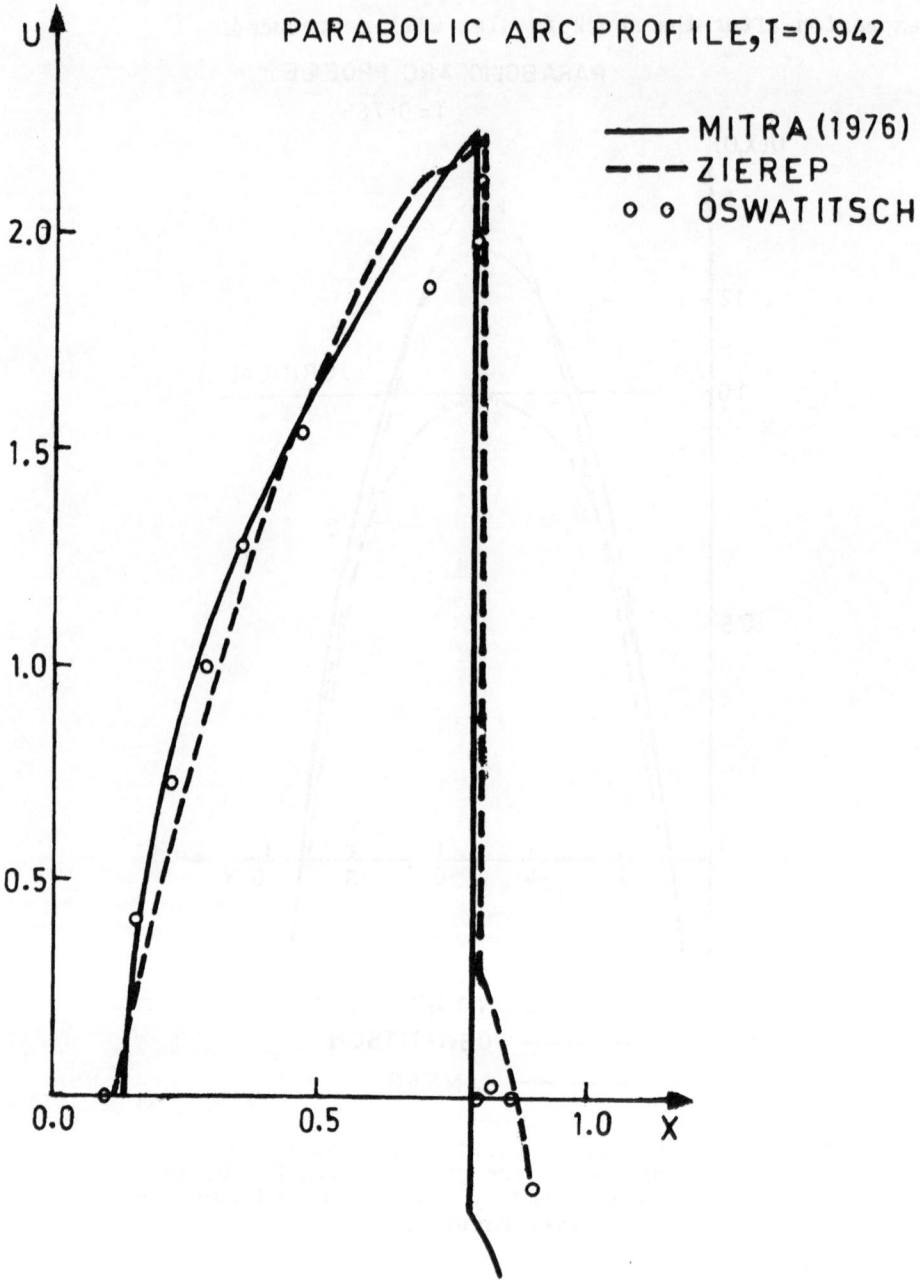

Fig.2.4   Reduced velocity distribution on a parabolic
arc profile with  T = 0.942, as computed by
Oswatitsch and comparison with other results.

We first note some useful observations made by Spreiter and Alksne regarding the nature of the solutions of the simplified Oswatitsch equation (14), which are also valid for the two dimensional equation (1). For this purpose, we put

$$J(U_0) = \int_0^1 \frac{U_0^2(\varepsilon)}{b} \; B(\frac{\varepsilon - X}{b}) \; d\varepsilon , \tag{17}$$

so that Eq.(14) may be solved as

$$U_0 = 1 \pm (1 - 2U_{P_0} + J)^{1/2} = 1 \pm (J - L)^{1/2}, \tag{18}$$

where

$$L = 2U_{P_0} - 1. \tag{19}$$

Thus, for obtaining real values of $U_0$ , it is necessary to have $J \geqslant L$. If in addition, the negative sign be selected in Eq.(18), then $U_0 < 1$, so that the flow is subsonic. On the other hand the positive sign in Eq.(18) yields $U_0 > 1$ , and the flow is supersonic. If $J = L$, so that $U_0 = 1$, the flow is sonic. Consequently, it may be concluded that the change in sign of the radical corresponds to a continuous change from subsonic velocities to supersonic velocities and vice-versa. If on the other hand, the sign of the radical changes discontinuously, a jump discontinuity in velocity occurs,  so that shocks appear. However, a shock can appear only in a supersonic flow, and therefore such a change in the sign of the radical in Eq.(18) can take place in the sequence positive to negative and not vicé-versa.

To compute the solution numerically, the domain of integration is subdivided into a number of subintervals, in each of which the integrand is assumed to be a constant (Fig.2.5). If $l_i$ be the length of the i-th subinterval whose mid-point is $\varepsilon_i$ and $U_{0i}$ be the value of $U_0$ at $\varepsilon_i$ , then this subinterval contributes to $J$ an amount:

$$\frac{1}{2} U_{0i}^2 \; \bar{B} \; (\frac{X - \varepsilon_i}{l_i} \; , \; \frac{2l_i}{b_i}) = \frac{U_{0i}^2}{2 \; b_i} \int_{\varepsilon_i - \frac{l_i}{2}}^{\varepsilon_i + \frac{l_i}{2}} B(\frac{\varepsilon - X}{b_i}) \; d\varepsilon. \tag{20}$$

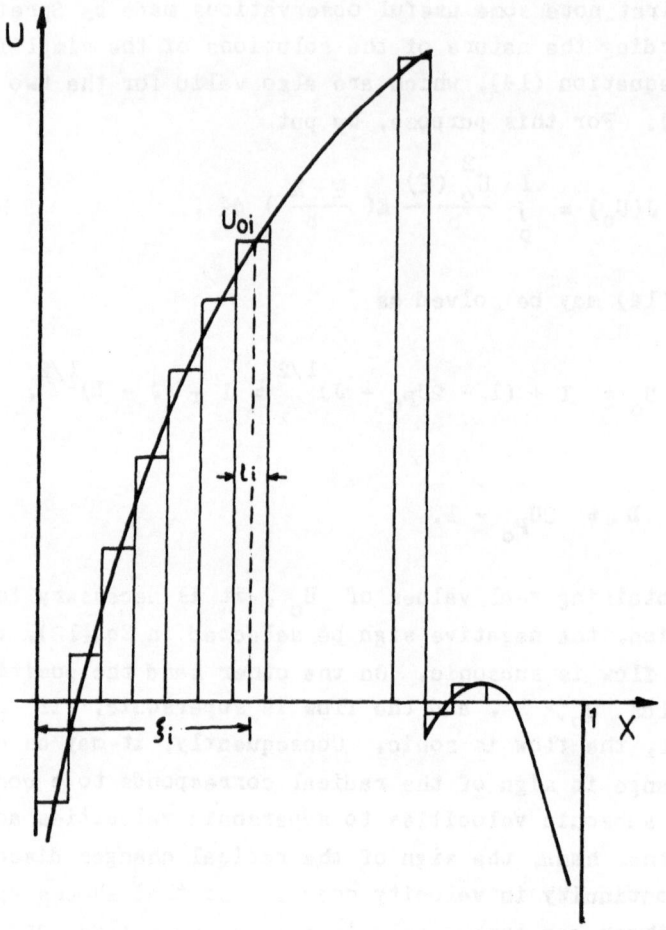

Fig.2.5     Domain of integration for numerical
            computation

The integration in Eq.(20) may be carried out in closed
form.  Performing the integration it follows from Eq.(20)

$$\frac{\pi}{4}\,\bar{g} = \frac{1}{12(1 + A^2)^4}\left\{ \frac{3\pi}{2}\,\frac{A}{|A|}\left[(1 + A^2)^4 - (1 + A^2)^2 + 8(1 + A^2) - 8\right] \right.$$

$$\left. + 12A(A^2-1)\log_e|A| - A(1+ A^2)\left[(1 + A^2)^2 + 12\right]\right\}$$

$$+ \frac{1}{12(1 + c^2)^4}\left\{ \frac{3\pi}{2}\,\frac{c}{|c|}\left[(1 + c^2)^4 - (1 + c^2)^2 + 8(1 + c^2) - 8\right] \right.$$

$$\left. + 12c(c^2 - 1)\log_e|c| - c(1 + c^2)\left[(1 + c^2)^2 + 12\right]\right\},$$

$$(21)$$

where

$$A = \frac{1}{4}\left(\frac{2\,l_i}{b_i}\right) + \frac{1}{2}\,\frac{X - \mathcal{E}_i}{l_i}\cdot\frac{2\,l_i}{b_i},$$

$$C = \frac{1}{4}\left(\frac{2\,l_i}{b_i}\right) - \frac{1}{2}\,\frac{X - \mathcal{E}_i}{l_i}\cdot\frac{2\,l_i}{b_i}. \qquad (22)$$

Consequently, the integral $J$ may be expressed as the following sum

$$J = 2\int_0^1 \frac{U_o^2}{2b}\,\bar{g}\left(\frac{\mathcal{E} - X}{b}\right)d\mathcal{E}$$

$$= 2\sum_{i=1}^{n} \frac{U_{oi}^2}{2}\,\bar{g}\left(\frac{X - \mathcal{E}_i}{l_i}, \frac{2\,l_i}{b_i}\right). \qquad (23)$$

Solution is computed by iteration according to the scheme

$$\overset{v}{U}_o = 1 \pm \left[1 - 2\,U_{P_o} + J(\overset{v-1}{U}_o)\right]^{1/2}, \qquad v = 1,2,3, \ldots,$$

$$\overset{o}{U}_o = U_{P_o} \qquad (24)$$

the positive or the negative sign of the radical being chosen as explained above, the superscript $v$ denoting the value of the solution at the $v$-th step of iteration.

For purely subsonic flow, the above scheme converges very rapidly. Further, when the initial distribution of $U_0$ along the chord is that corresponding to a local supersonic region terminated by a normal shock at $X = X_0$ , the iterations converge rapidly to a transonic flow with shock. However, no shock-free supercritical transonic solution could be found. Results presented for parabolic arc profiles, which may be found for example in Ferrari and Tricomi (1968) indicate hardly any quantitative improvement over the results of Oswatitsch. Failure to obtain shock-free supercritical solution indicate that they are also qualitatively unsatisfactory.

## 2.4  Method of Zierep

Since the method of Zierep has been reported in full details in the book of Zierep (1966), we present here only the essential steps of the procedure. For further details the book of Zierep (1966) and the paper Zierep (1962) may be consulted.

It is to be mentioned that Oswatitsch and his associates were concerned about the drawbacks of the substitution Eq.(9). Gullstrand (1951), (1952a,b) modified the work of Oswatitsch to the extent that a series substitution was used in place of Eq.(9). He also extended it to the case of small incidence, as also to the case of sonic free-stream conditions, discussed in section 2.5.

In an attempt to improve the substitution Eq.(9), Oswatitsch and Zierep (1962) and Zierep (1962), (1966) reformulated the problem and established the following two dimensional non-linear singular integral equation, now referred (Schubert and Schleiff 1969) to as the integral equation of Zierep:

$$U(X, Y) - \frac{1}{2} U^2(X, Y) - U_p(X, Y)$$

$$= \frac{1}{2\pi} \int_{-\infty}^{\infty} \int_{-\infty}^{\infty} U(\mathcal{E}, \eta) \frac{\partial U(\mathcal{E}, \eta)}{\partial \eta} \frac{\eta - Y}{(\mathcal{E} - X)^2 + (\eta - Y)^2} d\mathcal{E} d\eta ,$$

$$(25)$$

which is equivalent to the integral equation of Oswatitsch, Eq.(1), and may be derived from it by means of an integration by parts with respect to $\eta$ and using Oswatitsch principal value definition.

Oswatitsch and Zierep observed that the essential contribution to the double integral in Eq.(25) comes from the value of $U \frac{\partial U}{\partial \eta}$ near the body $\eta \approx 0$. In view of the irrotationality condition $\frac{dU}{d\eta}$ is related to the prescribed values of $\frac{dV_o}{d\mathcal{E}}$ on the profile, so that the double integral in Eq.(25) is essentially a linear function of the velocity.

To obtain a better approximation of the velocity field in terms of that on the profile axis, they used the following infinite series of singular integrals

$$U(X, Y) = \int_{-\infty}^{\infty} q_0(\mathcal{E}) \frac{Y}{(\mathcal{E}-X)^2 + Y^2} \, d\mathcal{E} + \int_{-\infty}^{\infty} q_1(\mathcal{E}) \, Y \frac{\partial}{\partial Y} \frac{Y}{(\mathcal{E}-X)^2 + Y^2} \, d\mathcal{E} + \ldots$$

$$+ \int_{-\infty}^{\infty} q_K(\mathcal{E}) \frac{Y^K}{K!} \frac{\partial^K}{\partial Y^K} \frac{Y}{(\mathcal{E} - X)^2 + Y^2} \, d\mathcal{E} + \ldots \,, \tag{26}$$

where $q_K$ denotes unknown velocity distribution. It is possible to express them in terms of the values on the profile axis as shown in Zierep (1966):

$$q_0(X) = \frac{1}{\pi} U(X, 0),$$

$$q_1(X) = \frac{1}{\pi} \left[ U_P(X, 0) - U(X, 0) \right],$$

$$q_2(X) = -\frac{2}{\pi} \left[ U_P - U + \frac{1}{4} U^2 \right], \tag{27}$$

$$\vdots$$

Substituting Eq. (26) in Eq. (25), using Eq.(27) and putting $Y = 0$ in it, two rather tedious integrations may be performed. Neglecting a rest term which is small, it may be put, after a little simplifications to the form

$$\frac{d}{dX} (U - \frac{1}{2} U^2 - U_P) = \frac{1}{2\pi} \int_{-\infty}^{\infty} \left\{ \frac{V_0(s)-V_0(X)}{s - X} \cdot \frac{\Phi(s) - \Phi(X)}{s - X} \right\} \frac{ds}{s - X}$$

$$= \overline{G} (X). \tag{28}$$

Eq.(28) is now solved by iteration using the scheme

$$U_n(X, 0) = U_P(X, 0) + \frac{1}{2} U^2_{n-1}(X, 0) + \int_{-\infty}^{X} \bar{G}_{n-1}(t) \, dt,$$

$$n = 1, 2, \ldots , \tag{29}$$

which is found to be convergent for supercritical transonic flow.

The solutions have been computed for a number of profile shapes like the wedge profile, parabolic arc profile and NACA 0012 profile. If the starting solution be shock-free the iteration converges to a shock-free supercritical solution. A solution with shock may be computed by starting with a solution containing a shock discontinuity, the position of the shock changing from step to step. For computation of shock, let $X = X_s$ denote the position of the shock in the n-th iteration step of computation and let the values before and after the shock be denoted by the superscript '-' and '+' respectively. Then, since the mass flux density, which equals $( U - \frac{1}{2} U^2)$ in this approximation, is conserved, we have at any step

$$U^+ - \frac{1}{2} U^{+2} = U^- - \frac{1}{2} U^{-2} , \tag{30a}$$

which delivers (for $U^+ \neq U^-$),

$$U^+ + U^- = 2. \tag{30b}$$

The iteration scheme delivers

$$U^+_n (X_s, 0) = U^+_P(X_s, 0) + \frac{1}{2} U^{+2}_{n-1} (X_s, 0) - J^+ \tag{31a}$$

$$U^-_n (X_s, 0) = U^-_P(X_s, 0) + \frac{1}{2} U^{-2}_{n-1}(X_s, 0) - J^-. \tag{31b}$$

In view of the fact that the linearized solution and the double integral, denoted by $J$ are continuous across the shock, it follows on addition of Eqs. (31) and using Eq.(30b)

$$2 = 2 U_P(X_s, 0) + \frac{1}{2} \left[ U^{+2}_{n-1} (X_s, 0) + U^{-2}_{n-1}(X_s, 0) \right] - 2J , \tag{32}$$

which is the equation determining the shock position in any step when that in the previous step is known. Initial guess of the shock position is taken to be the decelerating sonic point on the profile.

For a parabolic arc profile with reduced thickness ratio
T = 0.785, a shock-free supercritical solution as well as a solution
with shock could be found. For higher values of the reduced thickness
ratio, only the flow with shock exists. Generally, satisfactory
results have been found by this procedure of Zierep. Unlike the
earlier works, an electronic digital computer had to be used to manage
the great amount of computation needed. The method has not been extended to the case of lifting profile or to three dimensional problems.

The solution of Zierep for a parabolic arc profile has been
compared with that of Oswatitsch and Mitra (1976) in Fig.2.4. It will
be seen in Chapter III that the shock position determined according to
Mitra (1976) is in excellent agreement with those of modern finite-
difference relaxation procedures delivering results of high accuracy.
So, from Fig.2.4 it may be concluded that the shock position determined by Zierep is fairly good. Further, the critical value of the
reduced thickness ratio for a parabolic arc profile according to
Zierep is approximately T = 0.66, which is in excellent agreement with
recent numerical computations of Niyogi and Das (unpublished), who used
the full Oswatitsch equation (1).

Finally, we mention that for purely subsonic flow, Gretler
(1965) used integral equation formulation in developing his well-known
second order accurate method for computing flow past a moderately thick
profile at non-zero incidence.

2.5  Modification and extensions of Gullstrand

Immediately after the publication of the work of Oswatitsch
(1950), Gullstrand modified and extended the method in a number of
works (Gullstrand 1951, 1952a,b). In the case of flow past a thin
symmetric profile at zero incidence, Gullstrand (1951) used a series
substitution in place of Eq.(9) and solved the resulting equation
approximately.

Gullstrand (1952a) extended the integral equation formulation
to the case of sonic free stream for a thin symmetrical profile at
zero incidence including weak shocks. Applying Green's theorem to
the region shown in Fig.2.6, where $S_4$ and $S_5$ are two arbitrary
curves toward the trailing edge, Gullstrand (1952a) established the
integral equation

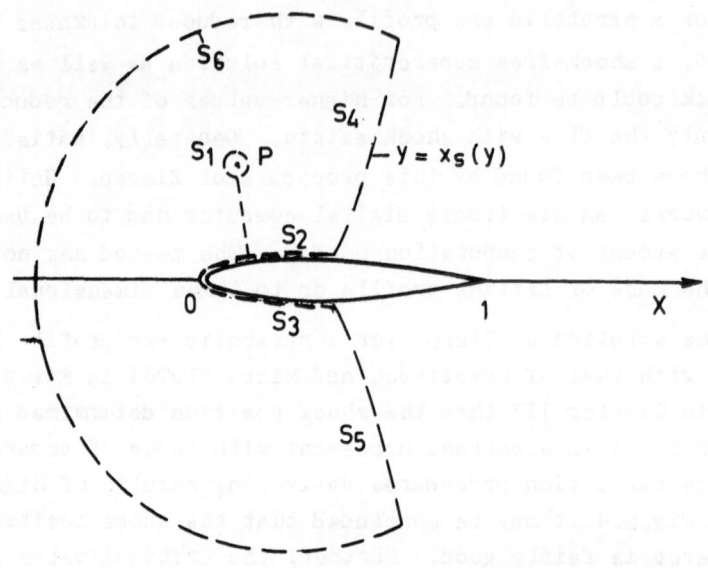

**Fig.2.6**    Domain of integration for sonic
free-stream condition

$$U(X,\ Y) - \frac{1}{2} U^2(X,\ Y) = U_L(X,\ Y;\ X_S) + \frac{1}{2\pi} \int_{-\infty}^{X_S} \int_{-\infty}^{\infty} \frac{1}{2} U^2(\xi,\ \eta)\ \Psi_{\xi\xi}\ d\xi d\eta$$

$$- \frac{1}{2\pi} \int_{-\infty}^{\infty} \left\{ U\ \Psi_\eta\ \cos\sigma - (U - \frac{1}{2} U^2)\ \Psi_\xi\ \sin\sigma + V\ \Psi_\xi \right\}_{\xi = X_S(\eta)} \cosec\ \sigma\ d\eta,$$

$$(33)$$

where the function $\Psi$ is defined in Eq.(1.91) and $\sigma$ denotes the
angle made by the tangent to the curve $S_4$ with the x-axis. Here
$U_L(X,\ Y;\ X_S)$ denotes the linearized solution

$$U_L(X,\ Y;\ X_S) = \frac{1}{\pi} \int_0^{X_S} V_0(\xi)\ \frac{\xi - X}{(\xi - X)^2 + Y^2}\ d\xi. \qquad (34)$$

Choosing the boundary $S_4$ parallel to the Y-axis and applying the
integral equation (33) to the domain in front of the straight line,
the simplified form, for $Y = 0$ is obtained as

$$U_o(X) = U_{P_o}(X ; X_S) + \frac{1}{\pi} \int_0^{\infty} \int_0^{X_S} \frac{1}{2} U^2(\xi, \eta) \; \Psi_{\xi\xi} \; d\xi d\eta$$

$$+ \frac{1}{\pi} \int_0^{\infty} \left[ U(\xi, \eta) - \frac{1}{2} U^2(\xi, \eta) - V(\xi, \eta) \Psi_{\eta} \right]_{\xi=X_S} d\eta .$$

$$(35)$$

It is convenient to choose the value of $X_S$ corresponding to the maximum thickness of the aerofoil, since the velocity at this point is certainly supersonic. For solving this equation Gullstrand (1952a) used series substitutions for $U(X, Y)$ and $V(X, Y)$ in terms of that on the profile axis. The remaining integral equation is then solved iteratively, performing the single integration numerically. The velocity distribution around a NACA 66-006 at $M_{\infty} = 1$ is presented in Fig.2.7, where $U_1$ denotes the starting solution.

NACA 66-006, $M_{\infty} \approx 1$

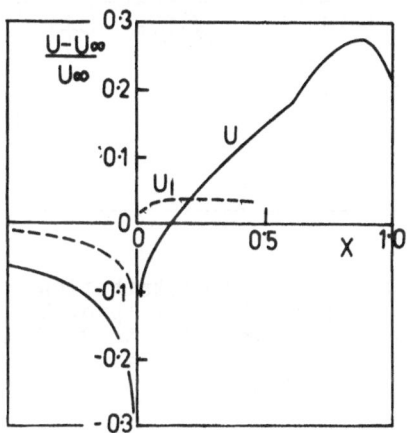

Fig. 2.7   The velocity distribution around a NACA 66-006 at $M_{\infty} = 1$ , after Gullstrand (1952a). $U_1$ starting solution

Gullstrand (1952b) extended the integral equation formulation to the case of a lifting profile and established the corresponding integral equations by application of Green's theorem. These integral equations are particularly simple if the additional velocity due to the lift is antisymmetric with respect to the chord of the aerofoil and if the angle of attack is small compared to the thickness of the aerofoil. In Nørstrud's first formulation, section 1.7.3, this corresponds to the assumption $\Phi^-$ small compared to $\Phi^+$ . Then the quadratic terms $\Phi_\zeta^- \Phi_{\zeta\zeta}^-$ may be neglected compared to the remaining terms in the equation for $\Phi_x^+$ Eq.(1.87a) and one obtains the trivial solution $\Phi^+ \equiv 0$.

The Mach number distribution at a 6 per cent symmetrical parabolic arc profile at zero lift for different free-stream Mach numbers as computed by Gullstrand (1952b) have been presented in Fig.2.8.

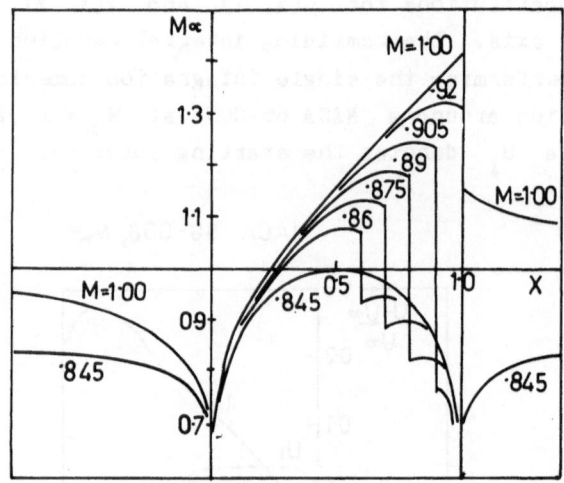

Fig. 2.8    Velocity distribution for a 6 per-cent
            parabolic arc  profile at zero lift
            corresponding to different free-stream
            Mach numbers as computed by Gullstrand
            (1952b)

The figure clearly shows choking of the flow with increasing   free-stream Mach number, when the velocity distribution over the greater part of the profile remains almost unchanged. Deviations occur only in the neighbourhood of the trailing edge because the shock recedes downstream with increasing free-stream Mach numbers.

It should be mentioned that the method of Oswatitsch and its extensions by Gullstrand were undertaken in the precomputer days. For solving the nonlinear singular integral equations, a number of simplifying assumptions had to introduced in order to simplify numerical calculations.  In spite of these assumptions, the results

obtained are surprisingly good, demonstrating the sound physical under-
standing of the authors.  In the second decade of development of the
method beginning from 1968, the authors were supported by powerful
electronic digital computers and they were able to introduce significant improvements in the numerical methods of solution.

## 2.6  Axisymmetric theory of Keune-Oswatitsch

Keune and Oswatitsch (1955) extended the integral equation method
to the axisymmetric case of flow past a slender axisymmetric body with
pointed tip at zero incidence.  They established the basic integral
equation for the 'spatial influence', simplified the basic equation
under plausible assumptions and solved the equation approximately by
the integral-to-sum method.  The main features of the method are presented in this section

Under the assumption of small perturbation, the gasdynamic equation for a slender axisymmetric body with noses and tails in the form
of the apex of a cone, at high subsonic free-stream Mach number may be
put to the form

$$(1 - 2\lambda^2 U) \frac{\partial U}{\partial X} + \frac{\partial V}{\partial Y} + \frac{V}{Y} = 0, \tag{36a}$$

and the irrotationality condition as

$$\frac{\partial U}{\partial Y} - \frac{\partial V}{\partial X} = 0, \tag{36b}$$

where  U  and  V  are reduced velocity components and  X  and  Y ,
reduced coordinates related to their true values denoted by the corresponding lower case letters by

$$M_\infty < 1, \quad U = \frac{u - u_\infty}{u_\infty (1 - M_\infty^2)} , \quad V = \frac{v}{u_\infty (1 - M_\infty^2)^{3/2}} ,$$

$$X = x , \quad Y = y(1 - M_\infty^2)^{1/2}, \tag{37}$$

and  $\lambda^2$  is a function of the free-stream Mach number, which may be
written as

$$\lambda^2 = M_\infty^2 (1 + \frac{\gamma - 1}{2} M_\infty^2). \tag{38}$$

In the above notation  y  denotes the radial distance from the symmetry axis of the body, taken as x-axis.  The reduced cross-section of the body may be expressed as

$$Q(X) = \frac{1}{1 - M_\infty^2} \pi h^2(x),$$  (39)

where  $y = h(x)$  denotes the radius of the body of revolution.

The tangency boundary condition at the body may be expressed as

$$\lim_{Y \to 0} VY = \frac{1}{2\pi} \frac{dQ}{dX}.$$  (40)

Noting that the perturbation velocities are infinite at all points on the axis of the body, it appears convenient to represent the perturbation flow in terms of the spatial influence and the cross-sectional flow, as is known from flow over low-aspect ratio wings and the Oswatitsch law of equivalence (Oswatitsch 1960).  Then, the perturbation potential  $\phi(X, Y)$  of the cross-sectional flow satisfies the equation

$$\frac{\partial^2 \phi}{\partial Y^2} + \frac{1}{Y} \frac{\partial \phi}{\partial Y} = 0.$$  (41)

At the location of the sources  $Y = 0$ , both the velocity components of this cross-sectional flow are infinite.  This flow alone contributes to infinite perturbation velocities and determines the boundary condition on the body.  The remaining part determines the spatial influence (the terminology being taken over from low-aspect-ratio wing theory). Thus the perturbation potential  $\Phi(X, Y)$  of the nonlinear flow described by Eqs. (36) is represented as

$$\Phi(X, Y) = \phi(X, Y) + \bar{\phi}(X, Y),$$  (42)

where  $\bar{\phi}(X, Y)$  determines the spatial influence. The linear theory of flow over low-aspect-ratio wings and bodies of revolution (Keune (1952), Keune and Oswatitsch (1953) ) shows that in the subsonic and supersonic regions the transverse flow is independent of the Mach number and that it also remains unaltered in the transonic region.  Only the spatial influence varies with the Mach number.  The equivalence law of Oswatitsch (see e.g. Oswatitsch and Keune (1954) ) shows that the total Mach number influence in the non-linear theory operates

only on the spatial influence.

Substituting Eq.(42), with $\Phi_X = U$, $\Phi_Y = V$ in Eq.(36), it follows

$$\phi_{XX} + \bar{\phi}_{XX} + \bar{\phi}_{YY} + \frac{1}{Y}\bar{\phi}_Y = \lambda^2 \frac{\partial}{\partial X} \Phi_X^2 \ . \tag{43}$$

The corresponding spatial influence function of linear theory $\bar{\phi}_L(X, Y)$ satisfies

$$\phi_{XX} + \bar{\phi}_{LXX} + \bar{\phi}_{LYY} + \frac{1}{Y}\bar{\phi}_{LY} = 0 \ , \tag{44}$$

since in linearized flow the right hand side of Eq. (36a) vanishes. Subtracting Eq.(44) from Eq.(43),

$$( \bar{\phi} - \bar{\phi}_L)_{XX} + ( \bar{\phi} - \bar{\phi}_L)_{YY} + \frac{1}{Y}( \bar{\phi} - \bar{\phi}_L)_Y = \lambda^2 \frac{\partial}{\partial X} \Phi_X^2 \ ,$$

which on differentiation with respect to $X$ delivers the differential equation

$$(\bar{\phi}_X - \bar{\phi}_{LX})_{XX} + (\bar{\phi}_X - \bar{\phi}_{LX})_{YY} + \frac{1}{Y}(\bar{\phi}_X - \bar{\phi}_{LX})_Y = \lambda^2 \frac{\partial^2}{\partial X^2} \Phi_X^2 \ .$$

$$\tag{45}$$

The integral equation of Keune-Oswatitsch is derived by application of Green's theorem to Eq.(45). After some lengthy simplification and manipulation, follows the integral equation for the axisymmetric case, assuming straight normal shock:

$$\bar{\phi}_{X0}(X) - \bar{\phi}_{LX0}(X) = \frac{1}{4} \int_{\eta=a}^{\infty} \left[ \bar{\phi}_X(\hat{X}, \eta) \right] \frac{\hat{X} - X}{\{(\hat{X} - X)^2 + \eta^2\}^{3/2}} \, d\eta^2$$

$$+ \lambda^2 \left\{ -\frac{1}{4} \int_{\eta=a}^{\infty} \left[ \bar{\phi}_X^2(\hat{X}, \eta) \right] \frac{\hat{X} - X}{\{(\hat{X} - X)^2 + \eta^2\}^{3/2}} \, d\eta^2 \right.$$

$$+ \frac{1}{2} \int_{-\infty, X}^{\hat{X}, \infty} \Phi_X^2 (\xi, a) \frac{a^2}{((\xi - X)^2 + a^2)^{3/2}} \, d\xi$$

$$\left. + \frac{1}{2} \int_{\xi=-\infty, s}^{s, \infty} \int_{\eta=a}^{\infty} \Phi_X (\xi, \eta) \cdot \eta \Phi_{XY} (\xi, \eta) \frac{d\eta^2 \, d\xi}{((\xi-X)^2 + \eta^2)^{3/2}} \right\}$$

$$\tag{46}$$

PARABOLIC SPINDLE , $M_\alpha=0.85$, $\Lambda^2=0.271$

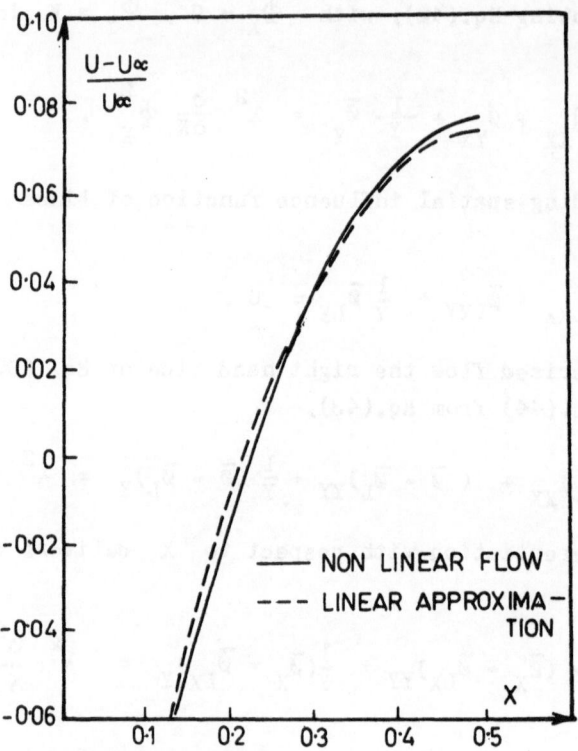

**Fig.2.9**   Perturbation velocity on a
parabolic spindle at $M_\infty = 0.85$,
according to Keune-Oswatitsch.

Here, the square bracket in the integrand of the single integral
represent the difference of the corresponding quantities across the
shock, $\hat{X}$ being the shock position, $\bar{\emptyset}_{X0}(X)$ and $\bar{\emptyset}_{LX0}(X)$ the
limiting values of $\bar{\emptyset}_{X}(X, Y)$ and $\bar{\emptyset}_{LX}(X, Y)$ as $Y$ approaches
0, a denotes the radius of the arbitrarily small cylinder used to
exclude the body axis from the domain of integration, and the abbre-
viation

$$\int_a^b f(x)\,dx + \int_b^c f(x)\,dx = \int_{a,b}^{b,c} f(x)\,dx , \qquad (47)$$

has been used. In this integral equation, the square of the velocity component is

$$\phi_X^2(X, Y) = \phi_X^2(X, Y) + 2 \phi_X(X, Y) \bar{\phi}_X(X, Y) + \bar{\phi}_X^2(X, Y).$$

(48)

The spatial influence function is unknown at all points in the field and its value on the body axis is sought. From physical considerations Keune-Oswatitsch assumed a variation in velocity with radial distance in the form

$$\bar{\phi}_X(X, Y) = \bar{\phi}_{X0}(X) \left[ (1 - Y^2)^2 + 3\omega(X) \, Y^2(1 - Y^2) \right],$$

(49)

where $\omega(X)$ is in general a function of the body. Using Eq.(49) it is then possible to simplify the basic integral equation (46), which is then solved by replacing the integrals approximately by sums. Computations have been carried out for a parabolic spindle in sub-critical flow. Figure 2.9 shows the perturbation velocity on a parabolic spindle with maximum thickness $T = 0.146$. Results of the linear approximation has also been presented for the sake of comparison. It appears that the linear and non-linear subcritical flow shows sufficient agreement upto high subsonic Mach numbers.

Chapter III

## Methods of Solution for Flow Past Symmetric Profiles
## at Zero Incidence

### 3.1 Introduction

The present and subsequent chapters are devoted to later developments in integral equation method for studying transonic aerodynamic problems. We begin with the simplest problem of shock-free supercritical transonic flow past a thin symmetric profile at zero incidence.
There was a long standing controversy (see e.g. Bers 1958) regarding
the existence of such flow fields. The celebrated non-existence
theorems of Morawetz (1956), could not resolve this controversy. Later,
the experimental works of Pearcy (1962) in England and the theoretical
works of Nieuwland (1968), as also the experimental works at NLR,
Netherlands (Spee and Uijlenhoet 1968) showed decisively the existence
of such flow fields. However, a mathematical proof of existence is
still lacking and we discuss the question of existence and uniqueness
of supercritical transonic flow as studied by the integral equation
method, in Chapter VII.

It is to be observed that even today, with very few exceptions,
all civilian transport aircrafts fly at Mach numbers rather less than
unity and similar is the case with most military aircrafts on low flying attack missions. It was felt for a long time that flying in the
transonic range would lead to drag and cost reduction in flight. To
do this it would be necessary to be able to analyze shock-free transonic flow past airfoils and wings and to design supercritical airfoils.
The analysis part of the problem has been discussed in the present
chapter, whereas Chapter VI is devoted to the design problem. Flow
past a thin symmetric airfoil at zero incidence with shock, has been
discussed in the later parts of the present chapter.

### 3.2 Approximate analytical solutions for shock-free flow

We consider steady inviscid flow past a thin symmetric airfoil
at zero incidence, with high subsonic free-stream Mach number. As
shown in details in Chapter I, the small perturbation problem may be
formulated in terms of an integral equation, known as the integral

equation of Oswatitsch. For studying shock-free flow, the alternative form of the integral equation of Oswatitsch, Eq.(1.79) appears to be more convenient, which was presented first by Schubert and Schleiff (1969), and later derived independently by Niyogi (1972) (See Niyogi 1976) and Nixon and Hancock (1974). The equation is

$$U(X, Y) = U_P(X, Y) + \frac{1}{4}U^2(X, Y)$$

$$- \frac{1}{2\pi} \int_{-\infty}^{\infty} \int_{-\infty}^{\infty} \frac{1}{2} U^2(\xi, \eta) \frac{(\xi-X)^2 - (\eta-Y)^2}{\left[(\xi-X)^2 + (\eta-Y)^2\right]^2} \, d\xi \, d\eta,$$

$$(1)$$

where $U_P(X, Y)$ is the known linearized solution according to the theory of Prandtl given by Eq.(1.80), $U(X, Y)$ is the reduced velocity component parallel to the free-stream direction and $X$ and $Y$ are reduced coordinates defined by Eqs. (1.20), (using Oswatitsch reduction) and repeated here for ready reference:

$$M_\infty < 1, \quad U = (u - u_\infty)/(c^* - u_\infty), \quad V = v \Big/ \left[ (c^* - u_\infty)(1 - M_\infty^2)^{1/2} \right],$$

$$X = x, \quad Y = y(1 - M_\infty^2)^{1/2}. \qquad (2)$$

The singularity at $\xi = X$, $\eta = Y$ in the double integral in Eq.(1) has been excluded by a circle of small radius. It is to be noted that, Eq.(1) is valid for shock-free flow as well as for flows with weak shocks, not necessarily straight and normal. (However, if a curved shock is present, the curvature has to be small).

At moderate subsonic free-stream Mach numbers, the terms in $U^2(X, Y)$ on the right of Eq.(1) are small and may be neglected in comparison with the first term, so that the linearized solution $U = U_P$ is regained. Further, for subcritical free-stream Mach numbers, the value of the double integral is small, compared to the other terms. Neglecting the double integral in Eq.(1)

$$U = U_P + \frac{1}{4} U^2(X, Y), \qquad (3a)$$

which may be rewritten as

$$U_P = \frac{1}{2} \left[ U + (U - \frac{1}{2} U^2) \right]. \qquad (3b)$$

This shows that (Oswatitsch 1976) the linearized solution equals the average of the non-linear velocity and mass-flux density. Eq.(3a) provides a rough approximation for the non-linear velocity distribution. Solving it as a quadratic equation in U and rejecting the purely supersonic solution yields

$$U(X,\ Y)\ =\ 2\left[1\ -\ \left\{1\ -\ U_p(X,\ Y)\right\}^{1/2}\right],\qquad (3c)$$

and on the profile axis Y = 0,

$$U(X,\ 0)\ =\ 2\left[1\ -\ \left\{1\ -\ U_p(X,\ 0)\right\}^{1/2}\right].\qquad (3d)$$

Solution (3d) was found by Niyogi (unpublished) and studied by Kundu (1972) and Chakraborty (1974) and independently by Nixon-Hancock (1974). Comparison with more accurate solutions shows that it provides only a rough approximation to the transonic solution which is however better than the linearized solution $U_p(X,\ 0)$. The approximation becomes worse with increasing free-stream Mach number and the best results are obtained for subcritical flow.

For shock-free supercritical flow, an approximate analytical solution of good accuracy has been given by the present author. In 1969, the exact solution of the following linear two dimensional singular integral equation, whose kernel is identical with that of the integral equation of Oswatitsch was found. In fact, it was established that the solution of

$$\bar{Q}(x,y)\ -\ \frac{1}{2\pi}\int_{-\infty}^{\infty}\int_{-\infty}^{\infty}\bar{Q}(\xi,\ \eta)\frac{(\xi-X)^2-(\eta-y)^2}{\left[(\xi-x)^2+(\eta-y)^2\right]^2}\ d\xi\ d\eta\ =\ \bar{g}(x,y),$$

$$(4a)$$

where the known function $\bar{g}(x,y)$ is square integrable in Lebesgue sense in the Euclidian space $E_2$, is given by (Niyogi 1969):

$$\bar{Q}(x,y)\ =\ \frac{2}{\sqrt{3}}\bar{g}(x,y)\ +\ \frac{2}{\sqrt{3}\pi}\int_{-\infty}^{\infty}\int_{-\infty}^{\infty}\bar{g}(\xi,\eta)\frac{(\xi-x)^2-3(\eta-y)^2}{\left[(\xi-x)^2+3(\eta-y)^2\right]^2}\ d\xi d\eta,$$

$$(4b)$$

and the solution is unique there. Proof of the inversion formula (4) has been presented in appendix - Al.

Similar linear two dimensional singular integral equations with constant coefficients have been studied by Niyogi (1973), Niyogi and Mitra (1973) and three dimensional linear singular integral equations with kernel having dipole singularity by A.K. Niyogi (1976).

Now, we rewrite Eq.(1) as

$$U^2(X, Y) - \frac{1}{2\pi} \int\limits_{-\infty}^{\infty} \int\limits_{-\infty}^{\infty} U^2(\mathcal{E}, \eta) \frac{(\mathcal{E} - X)^2 - (\eta - Y)^2}{\left[ (\mathcal{E} - X)^2 + (\eta - Y)^2 \right]^2} d\mathcal{E} \, d\eta$$

$$= 2U(X, Y) + \frac{1}{2} U^2(X, Y) - 2U_p(X, Y). \tag{5}$$

Applying the inversion formula (4) to Eq.(5) yields on simplification

$$(\sqrt{3} - 1) U^2(X, Y) - 4U(X, Y) + 4\left[ U_p(X, Y) - J(X, Y) \right] = 0, \tag{6}$$

where $J(X, Y)$ represents the double integral

$$J(X, Y) \equiv \frac{1}{4\pi} \int\limits_{-\infty}^{\infty} \int\limits_{-\infty}^{\infty} \left[ 4U(\mathcal{E}, \eta) + U^2(\mathcal{E}, \eta) - 4U_p(\mathcal{E}, \eta) \right].$$

$$\frac{(\mathcal{E} - X)^2 - 3(\eta - Y)^2}{\left[ (\mathcal{E} - X)^2 + 3(\eta - Y)^2 \right]^2} d\mathcal{E} \, d\eta. \tag{7}$$

Solving Eq.(6) as a quadratic equation in $U(X, Y)$, it follows

$$U(X, Y) = (\sqrt{3} + 1)\left[ 1 \pm \left\{ 1 - (\sqrt{3} - 1)(U_p(X, Y) - J(X, Y)) \right\}^{1/2} \right] \tag{8}$$

Noting that the upper sign before the radical leads to purely supersonic flow, we have at the profile  $Y = 0$ , the formal transonic solution

$$U(X, 0) = (\sqrt{3} + 1) \left[ 1 - \left\{ 1 - (\sqrt{3} - 1)(U_p(X, 0) - J(X, 0)) \right\}^{1/2} \right].$$

$$(9)$$

The double integral  $J(X, 0)$  still contains the unknown field velocity distribution  $U(\xi, \eta)$ . However, for shock-free flow the quantity  $J(X, 0)$  is small compared to  $U_p(X, 0)$  as shown by Niyogi (1976) on physical grounds. Neglecting it, we obtain from Eq.(9) an approximate analytical solution

$$U(X, 0) = (\sqrt{3} + 1) \left[ 1 - \left\{ 1 - (\sqrt{3} - 1) U_p(X, 0) \right\}^{1/2} \right],$$

$$(10)$$

which is a very simple relation expressing supercritical transonic solution only in terms of the corresponding linearized solution $U_p(X, 0)$ . It is to be observed from Eq.(10) that no real solution exist for

$$U_p(X, 0) > (\sqrt{3} + 1) / 2 (\approx 1.366 \cdots ),$$ $$(11)$$

which is the upper limit for shock-free flow.

The reduced surface pressure coefficient is of much interest, which is obtained from Eqs.(2.7) and (10) as

$$c_p \Big]_{surface} = -2(\sqrt{3} + 1) \left[ 1 - \left\{ 1 - (\sqrt{3} - 1) U_p(X, 0) \right\}^{1/2} \right]$$

$$(12)$$

Approximate solution (10) has been applied to many profile shapes (Haldar 1972).

An estimate of the error term in Eq.(10) was given by Chakraborty (1974) which showed it to be generally small. For a

parabolic arc profile the agreement of the approximate solution
Eq.(10) with that of the finite-difference solution of Murman and Cole
(1971) is excellent, as shown in figure 3.1  computed by Sen (nee'
Mitra) (1976).

The simple and handy formula (10) has been applied to a number
of other profile shapes, including Nieuwland profiles (Niyogi and Das
1979) and compared with exact solutions.  Since the approximate solution (10) is expressed in terms of the corresponding linearized  Prandtl
solution    $U_p(X, 0)$, a second order edge correction was used to take
care of the blunt leading edge, given by

$$U_p(X, 0) \Big]_{cor.} = \Big[ U_p(X, 0) \Big]_{EV.} \Big/ \Big\{ 1 + \frac{1}{\beta^2} \left( \frac{dh}{dX} \right)^2 \Big\}^{1/2} ,$$

$$\beta = (1 - M_\infty^2)^{1/2} . \qquad (13)$$

Here $\Big[ U_p(X, 0) \Big]_{cor.}$    denotes the corrected solution and $\Big[ U_p(X, 0) \Big]_{EV.}$

the evaluated solution and   $h(X)$   denotes the actual profile shape.

For Nieuwland aerofoil sections, the profile shape is prescribed by a set of numerical data.  A straight forward computation
of the linearized solution   $U_p(X, 0)$    from Eq.(1.80) by numerical
differentiation and by a subsequent numerical integration leads to
loss of accuracy due to the presence of the singularity at   $\xi = X$
on the profile axis.  To obtain results of sufficient accuracy,
Niyogi and Das (1979) represented the profile shapes by means of cubic
splines (Ahlberg et al  1967).  Subsequently the simple integration
was carried out analytically in terms of the coefficients of the
spline, which are obtained by solving a system of linear algebraic
equations on an electronic digital computer, Burroughs  B 6700.
Thirty pivotal points were chosen on the profile axis and the necessary
CPU  time for computing a profile shape  is about  4 seconds.

The computed solution for the quasi-elliptical aerofoil sections
NLR 0.1200 - 0.7500 - 0.000, 0.1200 - 0.7000 - 0.000, 0.1100 - 0.7500 -
0.9000  are shown in figures 3.2, 3.3  and  3.4  respectively, where
the corresponding exact solution of Nieuwland (1967) as computed by

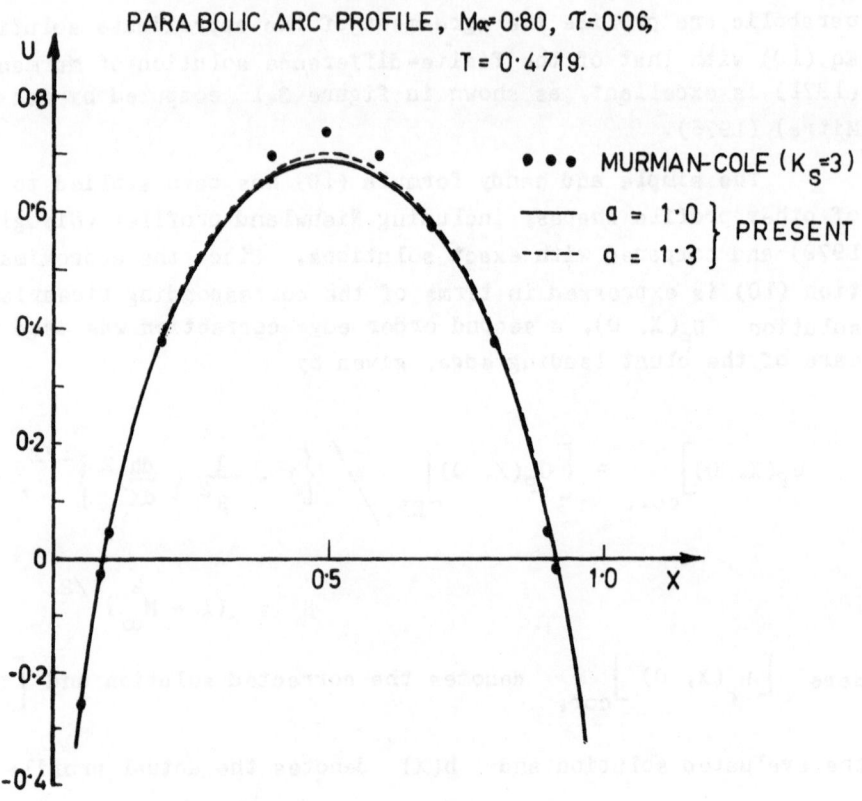

Fig.3.1    Velocity distribution for a parabolic arc profile
           in shock-free subcritical flow.  Comparison of
           solution (10) with finite-difference solution of
           Murman-Cole for the similarity parameter T = 0.4719.

Baurdoux and Boerstoel (1968) are also shown.  The solution for a NACA
0012 profile at $M_{\infty} = 0.72$ is shown in figure 3.5 and compared
with the finite-difference solution of the full potential equation by
Garabedian and Korn (Lock 1970).  Excellent agreement has been obtained
in all the cases.  It may be observed that the approximate solution
(10) delivers results with less than five per cent error.

      An attempt was made by Niyogi and Mitra (1973) to obtain an
improvement of the solution Eq.(10) by introducing a parameter in it.

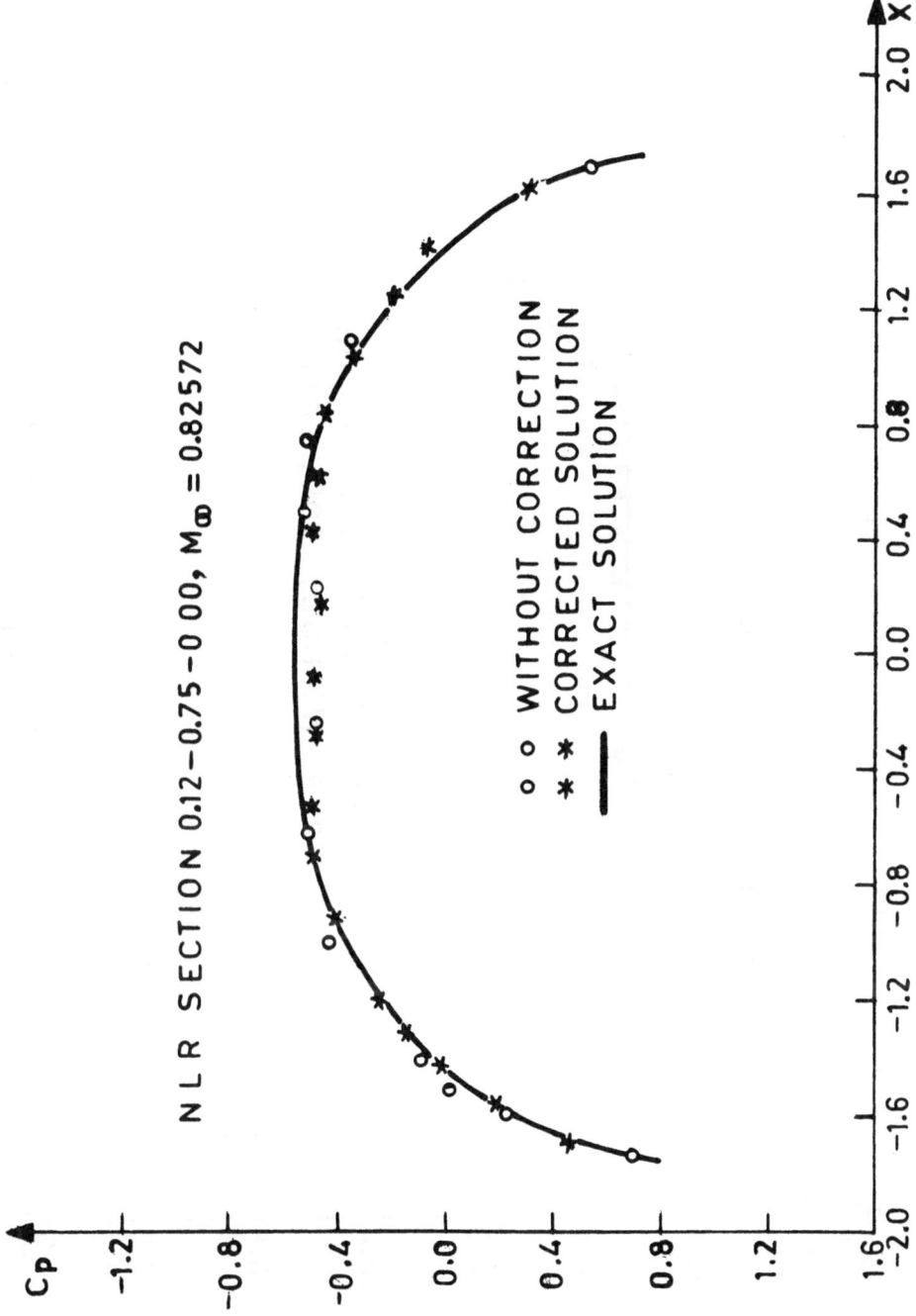

Fig.3.2 Surface pressure coefficient for Nieuwland aerofoil
section NLR 0.12 - 0.75 - 0.00, $M_\infty = 0.82572$, after
Niyogi and Das (1979). Comparison with the exact
solution

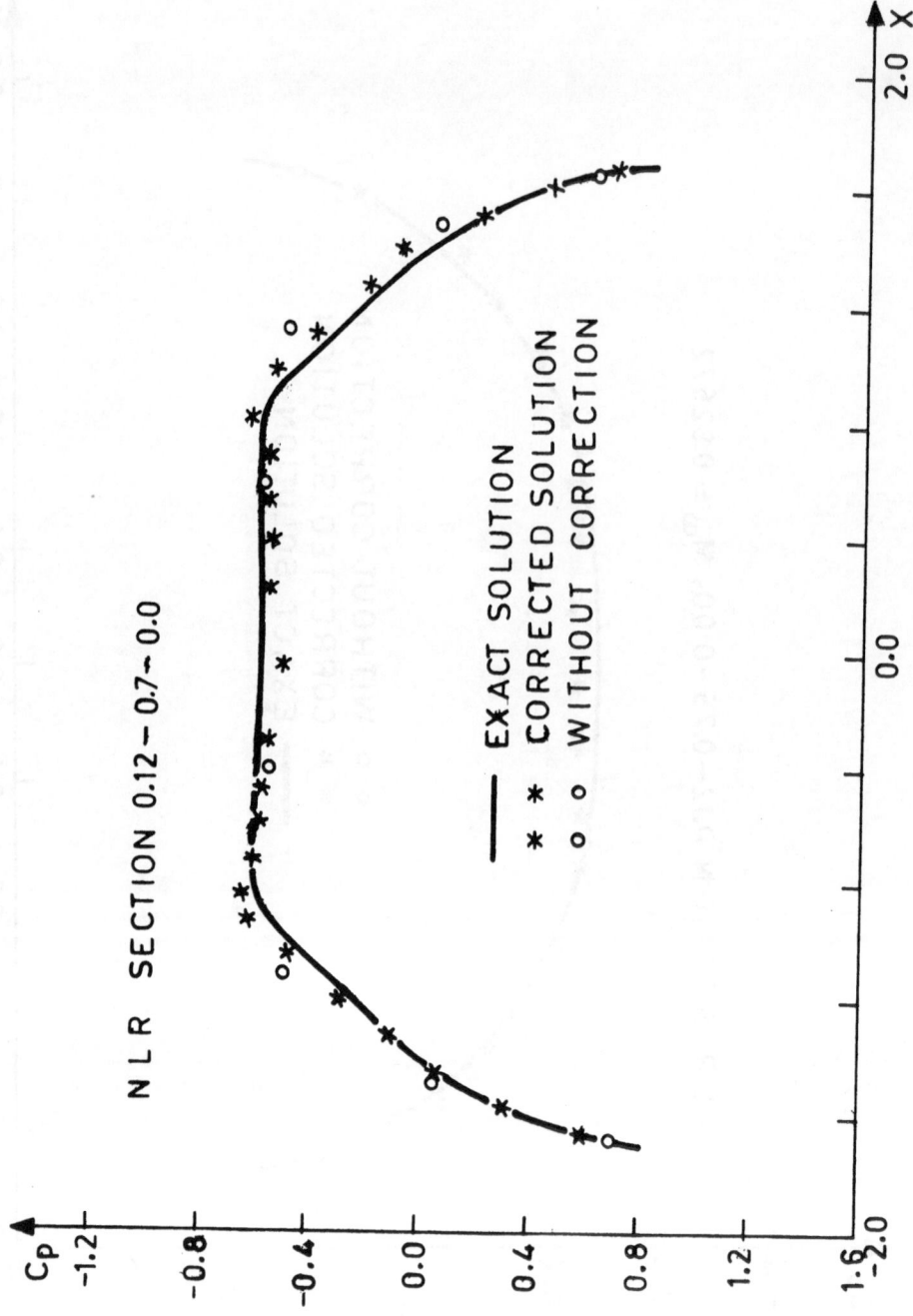

Fig.3.3 Surface pressure coefficient for Nieuwland aerofoil
        section  NLR 0.12 - 0.700 - 0.00, $M_\infty$ = 0.8257, after
        Niyogi and Das (1979).  Comparison with the exact
        solution.

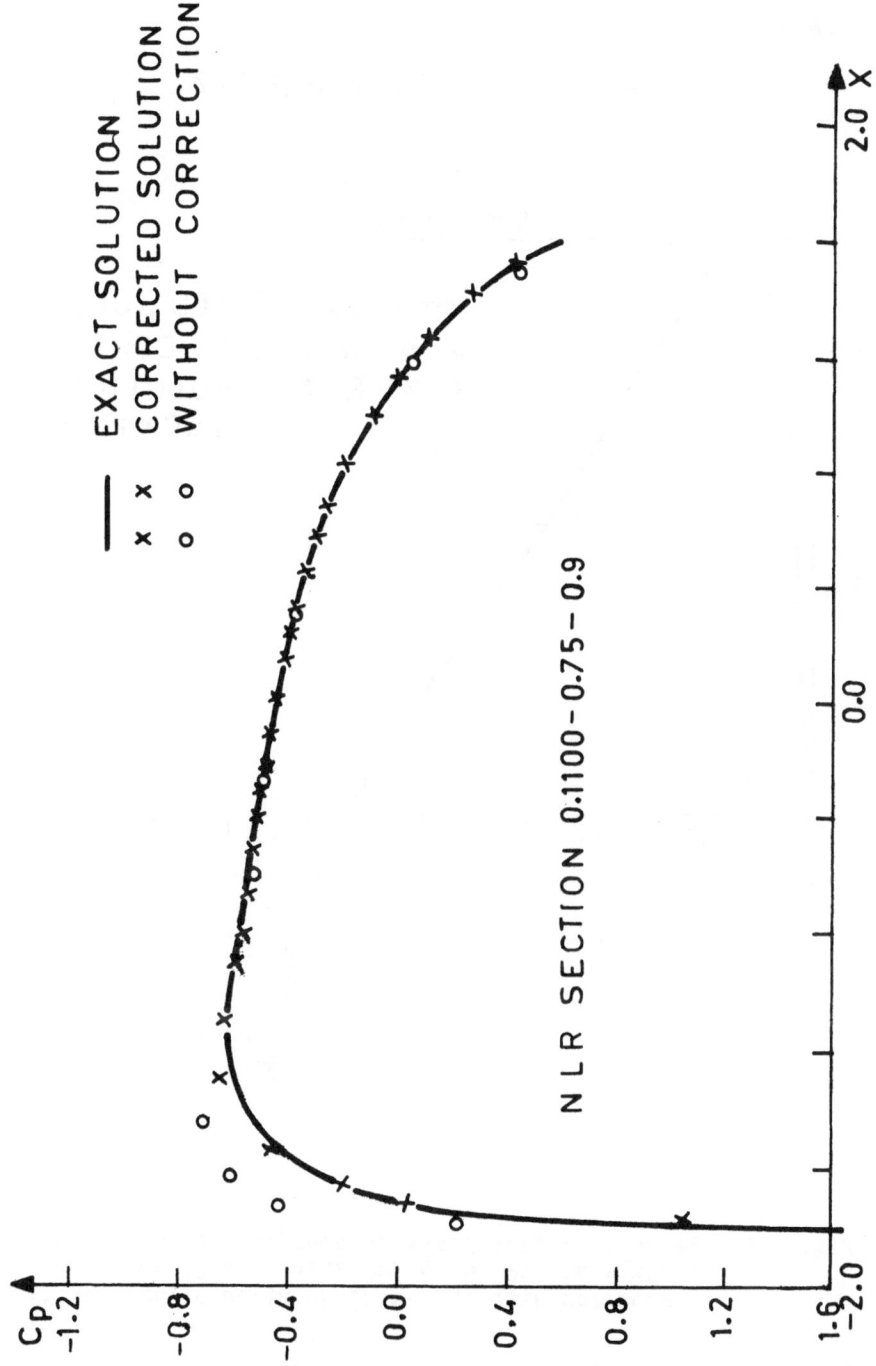

Fig.3.4   Surface pressure coefficient for Nieuwland aerofoil
          section    NLR 0.1100 - 0.75 - 0.9,   $M_\infty$ = 0.7861,
          after Niyogi and Das (1979).   Comparison with the
          exact solution.

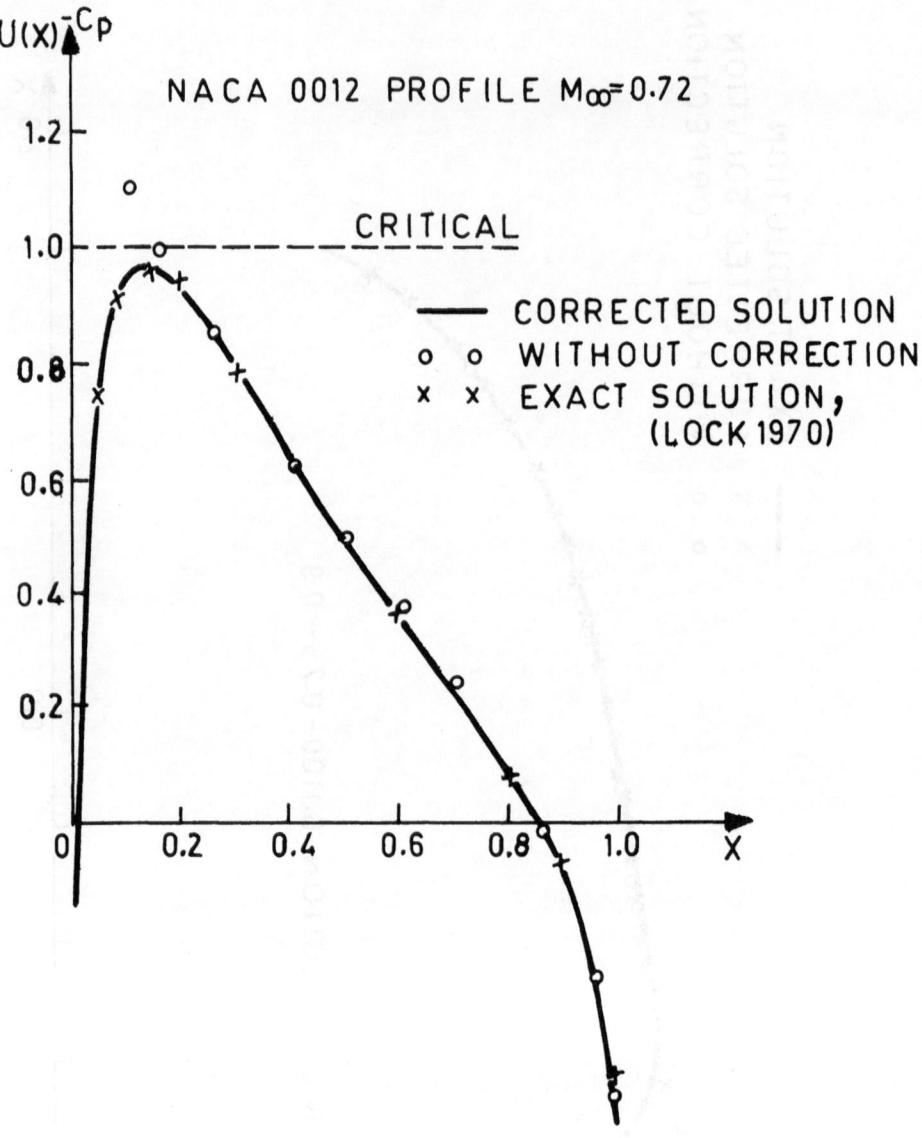

Fig. 3.5    Reduced surface pressure coefficient for a NACA 0012
            profile at Mₒₒ = 0.72, after Niyogi and Das (1979).
            Comparison with the exact solution after Lock (1970).

Noting that the alternative form of the integral equation of Oswatitsch may be rewritten in terms of a parameter  a  as

$$a\ U^2(X,Y) - \frac{1}{2\pi} \int_{-\infty}^{\infty} \int_{-\infty}^{\infty} U^2(\mathcal{E},\eta) \frac{(\mathcal{E}-X)^2 - (\eta-Y)^2}{\left[(\mathcal{E}-X)^2 + (\eta-Y)^2\right]^2} \,d\mathcal{E}\,d\eta$$

$$= 2\ U(X,Y) - 2\ U_p(X,Y) + (a - \frac{1}{2})\ U^2(X,Y), \qquad (14)$$

and applying the inversion formula (proved in appendix  A1 )

$$a\ \bar{Q}(x,y) - \frac{1}{2\pi} \int_{-\infty}^{\infty} \int_{-\infty}^{\infty} \bar{Q}(\mathcal{E},\eta) \frac{(\mathcal{E}-X)^2 - (\eta-y)^2}{\left[(\mathcal{E}-x)^2 + (\eta-y)^2\right]^2} \,d\mathcal{E}\,d\eta$$

$$= \bar{g}(x,y), \qquad (15a)$$

$$\bar{Q}(x,y) = \frac{2}{(4a^2-1)^{1/2}} \bar{g}(x,y)$$

$$+ \frac{2}{\pi(4a^2-1)^{1/2}} \int_{-\infty}^{\infty} \int_{-\infty}^{\infty} \bar{g}(\mathcal{E},\eta) \frac{(2a-1)(\mathcal{E}-x)^2-(2a+1)(\eta-y)^2}{\left[(2a-1)(\mathcal{E}-x)^2+(2a+1)(\eta-y)^2\right]^2} d\mathcal{E}d\eta,$$

$$\qquad (15b)$$

which is valid if the known function belongs to  $L_2(E_2)$  and  a  is a parameter greater than  $\frac{1}{2}$ , the following approximate solution was found:

$$U(X,0) = l(a)\left[1 - \left\{1 - \frac{2}{l(a)} U_p(X,0)\right\}^{1/2}\right]. \qquad (16)$$

Here  l(a)  is the parameter function

$$l(a) = 1 + (4a^2-1)^{1/2} \big/ (2a-1). \qquad (17)$$

In deriving the approximate solution Eq.(16) a singular double integral was neglected on physical grounds. However, it turns out that the optimum value of the parameter a is close to unity, for which value solution Eq.(10) is regained. The variation of the solution for two different values of the parameter a for a subcritical parabolic arc profile are shown in Fig.3.1 and compared with the finite-difference solution of Murman and Cole (1971) for the similarity parameter $K_S = 3$.

### 3.3  Iterative solutions of the simplified Oswatitsch equation

The encouraging results obtained from the approximate analytical solution Eq.(10) suggests its iterative improvement. Use of iterative methods in the transonic range needs some caution, because frequently they fail to converge in the supercritical range (Oswatitsch 1956, Chap. IX). In search of satisfactory iteration schemes, it is found that the direct iteration scheme (briefly DIS), put forward by Niyogi and Chakraborty (1979) is such a rapidly convergent scheme. The direct iteration scheme for the simplified Oswatitsch equation (2.14) is defined by

$$U_{n+1}(X,0) = U_P(X,0) + \frac{1}{2} U_n^2(X,0)$$

$$- \int_0^1 \frac{U_n^2(\mathcal{E},0)}{2b(\mathcal{E};U_n(\mathcal{E},0))} E \left[ \frac{\mathcal{E}-X}{b(\mathcal{E};U_n(\mathcal{E},0))} \right] d\mathcal{E},$$

$$n = 0,1,2,3, \ldots\ldots,$$

$$(18a)$$

with the starting solution $U_0(X,0)$ given by Eq.(10)

$$U_0(X,0) = (\sqrt{3}+1)\left[1 - \left\{1 - (\sqrt{3}-1)U_P(X,0)\right\}^{1/2}\right].$$

$$(18b)$$

To carry out the above computations, the linearized solution in reduced coordinates $U_P(X,0)$ has to be calculated first. For a parabolic arc profile defined by Eq.(2.15), using Eq.(2.2) and the tangency boundary condition Eqs.(2.3) - (2.5) we obtain by simple integration

$$U_p(X,\ 0)\ =\ \frac{4T}{\pi}\left[1\ +\ (X\ -\ \tfrac{1}{2})\ \ln\ \left|\frac{1\ -\ X}{X}\right|\ \right], \tag{19}$$

and for a NACA 0012 profile defined by Eq.(2.16)

$$U_p(X,\ 0)\ =\ \frac{T}{0.2\,\pi}\left[\ \frac{a_1}{2\sqrt{X}}\ \ln\left|\frac{1\ +\ \sqrt{X}}{1\ -\ \sqrt{X}}\right|\ -\ \left\{\frac{4}{3}a_5\ +\ \tfrac{1}{2}(3a_4+4a_5 X)\right.\right.$$

$$+\ (2a_3\ +\ 3a_4 X\ +\ 4a_5\ X^2)\ +\ (a_2+2a_3 X+3a_4\ X^2+4a_5 X^3\ )\ln\left|\frac{1-X}{X}\right|\Bigg],$$

$$\tag{20a}$$

where

$$a_1\ =\ 0.29690,\qquad a_2\ =\ -0.12600,\qquad a_3\ =\ -0.35160,$$

$$a_4\ =\ 0.28430,\qquad a_5\ =\ -0.10150. \tag{20b}$$

The singular integral in Eq.(18a) may be evaluated numerically, taking adequate care for the singularity at $\mathcal{E} = X$. The numerical integration scheme discussed in section 2.3, may also be used.

As explained in details in Niyogi and Chakraborty (1979), the direct iteration scheme Eqs.(18) converges rapidly to a shock-free supercritical flow in the range of reduced thickness ratio (which is a transonic similarity parameter) $0.596 < T \leqslant 0.64$ for a parabolic arc profile. Further, in course of computation it has been found that in the above range of reduced thickness ratio, a discontinuous starting solution smoothes out and converges to a shock-free supercritical flow, indicating stability of DIS against small disturbances. Moreover, for the following starting solutions

$$U_0\ =\ 2\left[1\ -\ \left\{1\ -\ U_p(X,\ 0)\right\}^{1/2}\ \right], \tag{21a}$$

$$U_o = U_p(X, 0), \qquad (21b)$$

$$U_o = 1, \qquad (21c)$$

DIS converges also to the same supercritical solution, correct upto two decimal places in only ten to fifteen iteration steps. For a NACA 0012 profile, computations show that $0.72 < T \leqslant 0.75$ is the range

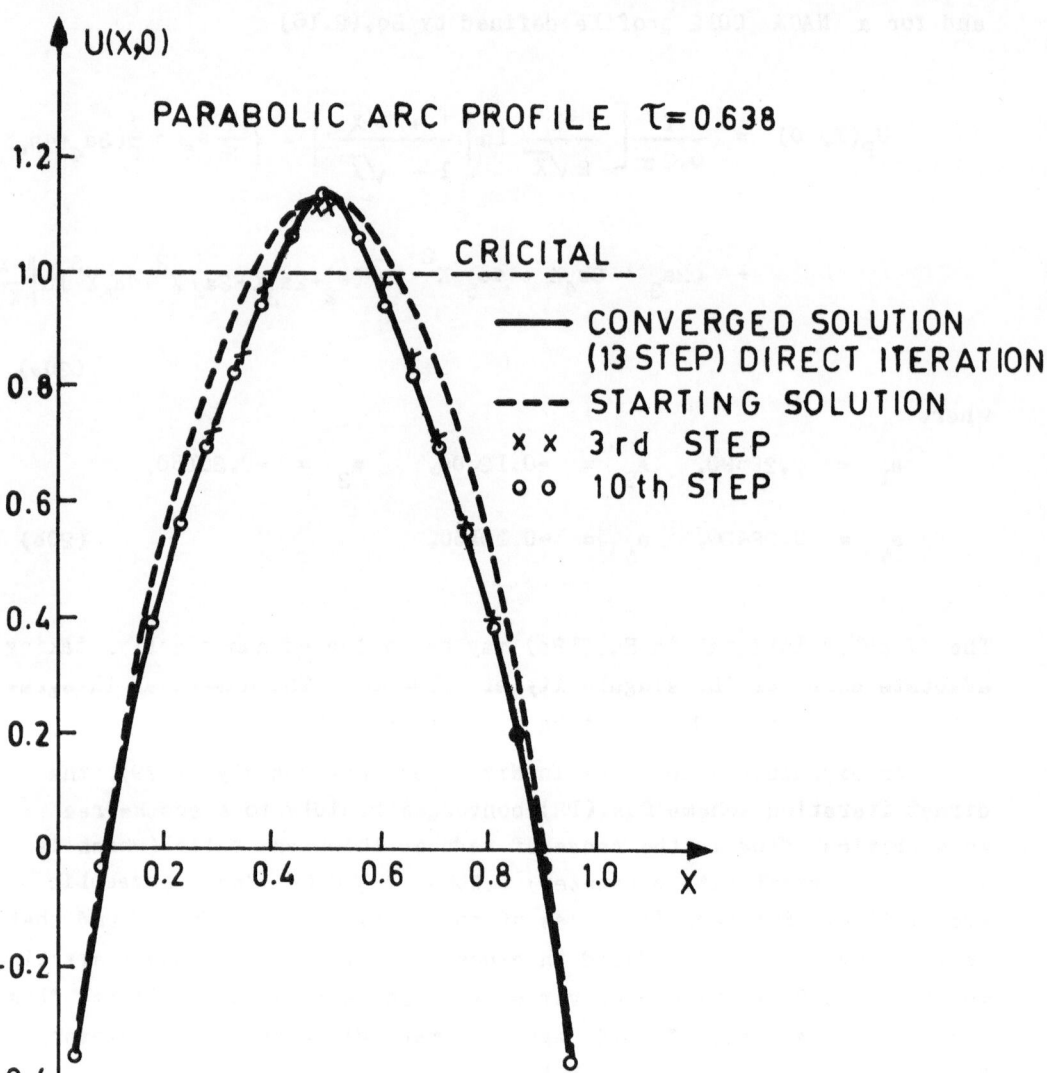

Fig.3.6  Reduced surface velocity distribution for a parabolic arc profile in supercritical flow by DIS T = 0.638.

in which shock-free supercritical solution exists.

The solution for a parabolic arc profile for T = 0.638 computed with forty pivotal points on the profile axis is shown in Fig.3.6. With Eq.(21a) as the starting solution, converged shock-free supercritical solution was obtained in 13 iteration steps, correct up to two decimal places, requiring about 30 seconds of CPU time on a Burroughs B6700 computer system. If the number of pivotal points be doubled, the number of steps needed for convergence is less than twentyfive for the same tolerance.

For a NACA 0012 profile with free-stream Mach number $M_\infty = 0.75$, the results of DIS has been compared with the corresponding results of Jameson (1974) and Nørstrud (1973a) in Fig.3.7.

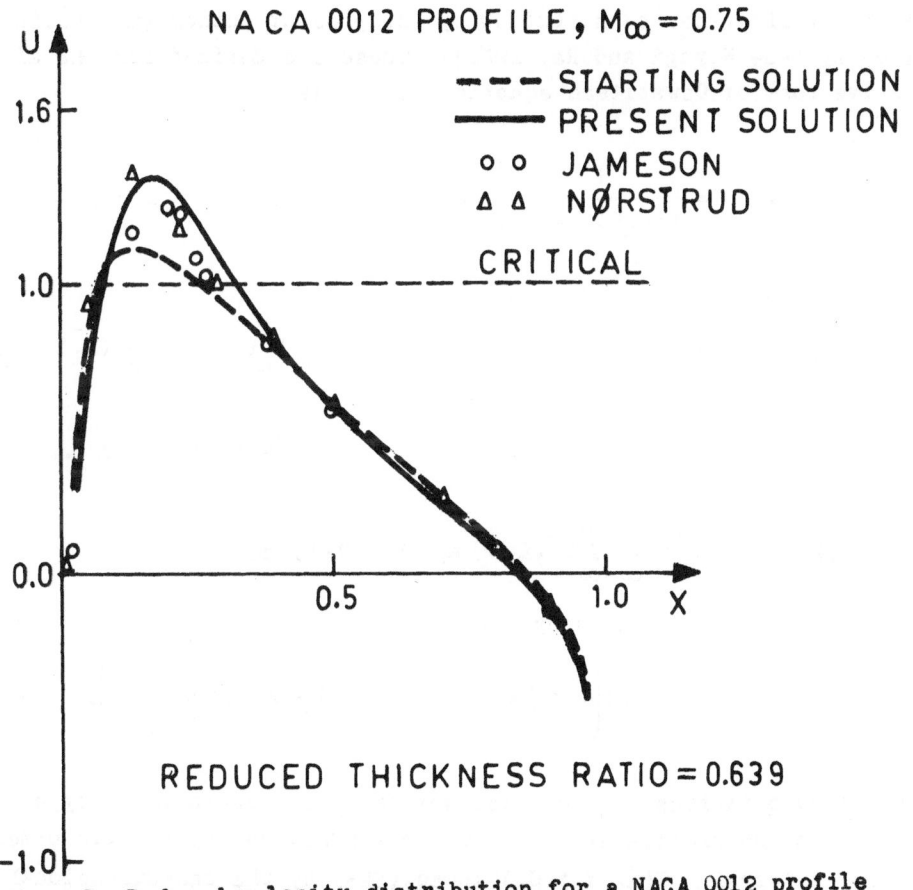

Fig.3.7 Reduced velocity distribution for a NACA 0012 profile at $M_\infty = 0.75$, by DIS. Comparison with the results of Jameson (1974) and Nørstrud (1973a).

The present solution, computed with forty pivotal points on the profile axis converges in only thirteen iteration steps correct up to two decimal places. The agreement is quite satisfactory, except for a small discrepancy near the pressure peak. However, this discrepancy cannot be removed even with double the number of pivotal points and seems to be inherent in the approximation of the double integral by the one-dimensional model, viz. the simplified Oswatitsch equation (2.14). No shock-free supercritical converged solution is found beyond the above mentioned ranges, and the solution diverges due to the formation of an expansion shock at the accelerating sonic point. To construct solutions with a shock discontinuity, some modifications have to be introduced in the DIS, which is discussed in section 3.5.

Other iteration schemes,which converge for shock-free super-critical flow, has been studied by the present author (unpublished) and Ray (see Niyogi and Ray 1981). These are defined for the alternative form of Oswatitsch equation (1.79) as

$$A) \quad U_{n+1}(X,Y) = \frac{4}{4-U_n(X,Y)}\left[U_P(X,Y) - \frac{1}{2}I(X,Y;U_n)\right], \qquad (22a)$$

$$B) \quad U_{n+1}(X,Y) = \frac{2}{2-U_n(X,Y)}\left[U_P(X,Y) - \frac{1}{4}U_n^2(X,Y) - \frac{1}{2}I(X,Y;U_n)\right],$$

$$n = 0, 1, 2, 3, \ldots\ldots,$$

where

$$I(X,Y;U) \equiv \frac{1}{2\pi}\int_{-\infty}^{\infty}\int_{-\infty}^{\infty} U^2(\mathcal{E},\eta)\,k_2(\mathcal{E}-X,\eta-Y)\,d\mathcal{E}\,d\eta, \qquad (22b)$$

the starting solution being taken as

$$U_0(X,Y) = (\sqrt{3}+1)\left[1 - \left\{1 - (\sqrt{3}-1)U_P(X,Y)\right\}^{1/2}\right]. \qquad (22c)$$

Actual computations were carried out for the reduced velocity distribution on the profile axis $Y = 0$, and approximating the two dimensional integral $I(X,Y;U_n)$ in Eq.(22a) by the one-dimensional approximate model of Oswatitsch discussed in section 2.2, the numerical computations being carried out as in the case of DIS. With these schemes also, rapid convergence to shock-free supercritical flow has been found, the rate of convergence being much faster than that of DIS. For example, whereas DIS requires about 13-15 iteration steps in the shock-free supercritical case for results correct up to 2-decimal places, it now takes only 7-8 steps for scheme (A) and 5-6 steps for scheme (B) for the same accuracy. Illustrative examples have not been presented here, as the final converged solutions come out to be identical with that of DIS.

Under the suggestion and supervision of the present author, a third iteration scheme, now known as the square-root iteration scheme was studied by Kundu (1972) and investigated further by Chakraborty (1974) and Niyogi and Chakraborty (1979). It is defined by

$$U_{n+1}(X, 0) = 2\left[1 - \left\{1 - U_p(X, 0) + \frac{1}{2}I(X, 0; U_n)\right\}^{1/2}\right],$$

$$n = 0,1,2, \ldots, \tag{23}$$

with the starting solution Eq.(21a), the double integral $I(X,0;U_n)$ being approximated and evaluated as explained above for the previously mentioned iteration schemes.

Computations carried out with the square-root iteration scheme indicate rapid convergence for subcritical flow, the iterations converging in only five steps correct up to two decimal places for a parabolic arc profile. However, for supercritical flows, the iteration scheme fails to converge near the leading and trailing edges and also near the maximum thickness of the profile, as a result of which the whole iteration diverges. It fails to converge for supercritical flow also with other starting solutions like Eq.(18b) or Eqs. (21).

The computational results for a parabolic arc profile of reduced thickness ratio $T = 0.625$, which is in the supercritical range has been presented in Fig.3.8, where the corresponding converged solution obtained by the DIS is also presented.

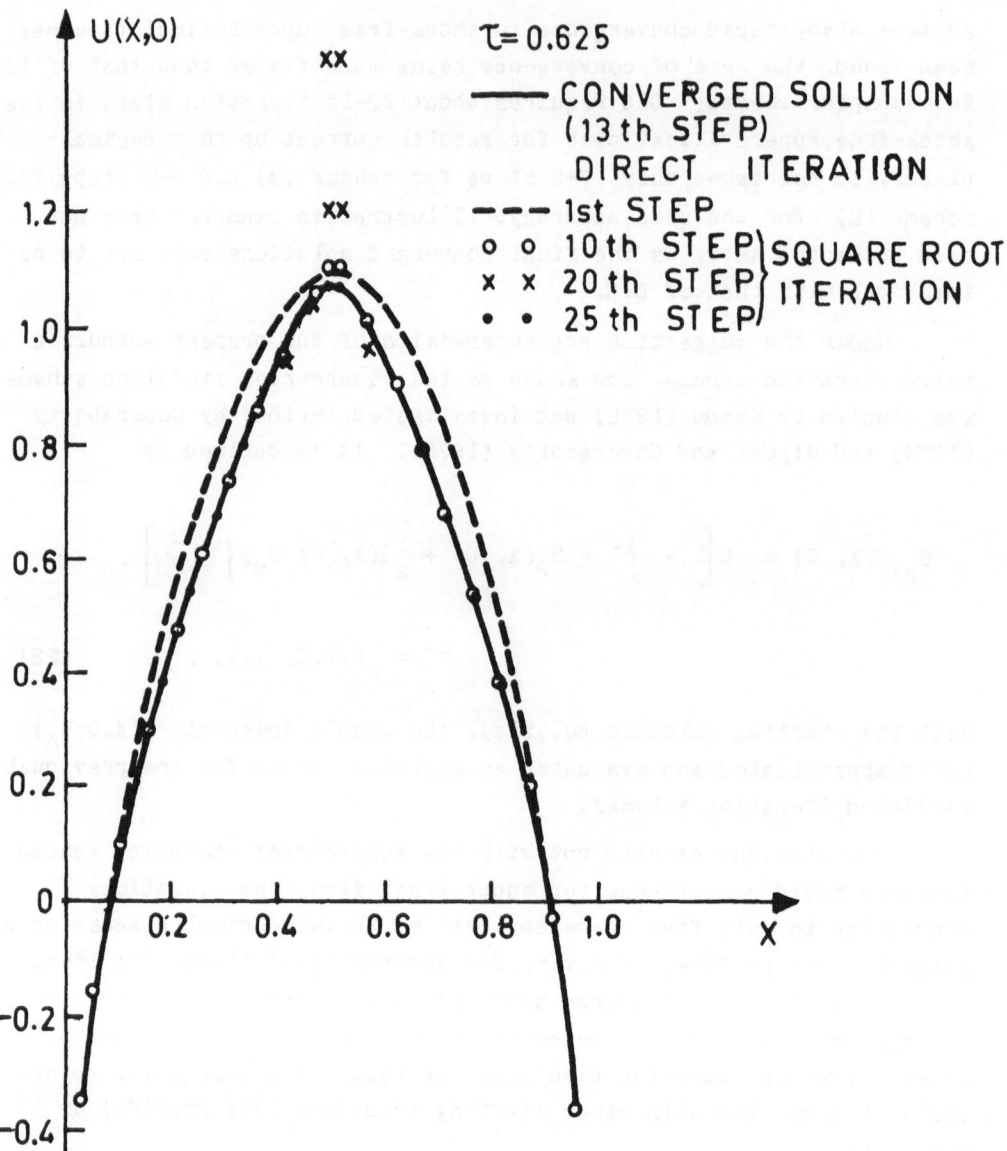

Fig.3.8 Divergence of the square-root iteration for supercritical flow past a parabolic arc profile with T = 0.625. Comparison with the DIS.

From computational results, it is observed, that for the square-root scheme there is an unsuccessful tendency to form an expansion shock at about $X = 0.47$ and a compression shock at $X = 0.53$, the flow remaining perfectly symmetrical about $X = 0.50$, and ultimately it diverges.

The square-root scheme with the starting solution Eq.(21a) was also used later, independently by Nixon and Hancock (1974) and subsequently evaluated by Nixon and Patel (1975). They computed only the first iteration step on the profile axis and claimed satisfactory convergence for a parabolic arc profile and a NACA 0012 profile, while no convergence was found by them for the Garabedian-Korn profiles or NPL 3111 profiles. However, since they did not carry out the iterations to a sufficient number of steps, their conclusions are not dependable.

Finally, we mention the iterative series proposed by Schubert and Schleiff (1969):

$$U = U_P + A \left\{ U_P^2 \right\} + A \left\{ 2 U_P A(U_P^2) \right\} + \ldots \ , \tag{24a}$$

where $\quad A \left\{ U^2 \right\} \quad$ represents the operator

$$A \left\{ U^2 \right\} \equiv \tfrac{1}{4} U^2 - \tfrac{1}{2} I(X, Y; U), \tag{24b}$$

the double integral $I(X, Y; U)$ being defined by Eq.(22b). The series converges to the unique exact solution of the alternative form of the Oswatitsch integral equation (1) for $U < 1$, that is for subcritical flow. Computations carried out with the above series (Chakraborty 1974, Niyogi and Chakraborty 1979), show that the terms of the Schubert-Schleiff series Eqs. (24) decrease rapidly for subcritical flow, so that the first three terms of the series are capable of delivering useful approximate results with less than 10 per cent error.

## 3.4 Approximate analytical solution for flow with shock

The approximate analytical solution for shock-free flow Eq.(10), has been extended to the case of flow with shock. This was carried out by Mitra (1976) under the suggestion and guidance of the present author. The main idea of the method is to determine a correction term $\bar{g}(x, y)$ to the perturbation potential $\phi_1(x, y)$ corresponding to the solution Eq.(10) expressed in terms of the true variables $x$ and $y$. The correction term $\bar{g}(x, y)$ will in general be small compared to the starting solution $\phi_1(x, y)$. We are particularly interested in $\bar{g}(x, y)$ possessing a jump discontinuity, at which position however the correction may not be small. Then with the definition

$$\phi(x, y) = \phi_1(x, y) + \bar{g}(x, y), \tag{25}$$

going over to the governing small perturbation partial differential equation (in non-reduced form) and carrying out an order of magnitude study of the different terms, as shown in details in Mitra (1976), it is found that $\bar{g}(x, y)$ satisfies the following second order non-linear equation

$$\frac{\partial}{\partial x}\left[(\phi_{1x} - \frac{1 - M_\infty^2}{\mu})\,\bar{g}_x + \frac{1}{2}\bar{g}_x^2\right] + \frac{3 - \sqrt{3}}{2}\,\phi_{1x}\,\phi_{1xx}$$

$$- \frac{(1 - M_\infty^2)^2}{\mu^2}\,\frac{\partial u_P}{\partial x} - \frac{1}{\mu}\,\phi_{1yy} = 0, \tag{26}$$

the suffixes denoting differentiation. Here $\mu$ is a Mach number function defined by

$$\mu = (c^* - u_\infty)\,/\,u_\infty = (1\,/\,M_\infty^*) - 1. \tag{27}$$

In deriving Eq.(26) terms of order higher than $t^2$, with $t = \tau^{2/3}$, $\tau$ being the thickness ratio of the profile, have been neglected. It is to be noted that in the transonic speed range, the u-perturbation is of the order $O(t)$. Integrating Eq.(26) and solving it as a quadratic equation in $\bar{g}_x$ follows the corrected solution in reduced coordinates

$$U(X, Y) = 1 \pm \left[ \left\{ U_1(X, Y) - 1 \right\}^2 - \frac{3 - \sqrt{3}}{2} \left\{ U_1^2(X, Y) - 1 \right\} \right.$$

$$\left. + 2 \left\{ U_p(X, Y) - U_p(X_1^*, Y) \right\} + 2 \int_{X_1^*}^{X} \frac{\partial V_1}{\partial Y} dX \right]^{1/2}, \qquad (28a)$$

where the starting solution $U_1(X, Y)$ is taken as

$$U_1(X, Y) = (\sqrt{3} + 1) \left[ 1 - \left\{ 1 - (\sqrt{3} - 1) U_p(X, Y) \right\}^{1/2} \right], \quad (28b)$$

and the constants and arbitrary function of integration being determined at the approximate accelerating sonic point $X = X_1^*$ on the
profile axis, it being assumed that at that point the true velocity
is the same as the approximating starting velocity. This condition
introduces some error, however not much, in view of the fact that the
starting solution is quite good near the accelerating sonic point. In
Eq.(28a) the upper and lower signs of the radical are to be taken
according as $U_1 \gtrless 1$.

The above solution has been computed for a number of profile
shapes including parabolic arc profile and NACA 0012 profiles and
compared with other results. Generally, satisfactory agreement have
been found. The present solution for a parabolic arc profile has been
compared with the finite difference solution of Murman and Cole (1971)
in Fig.3.9. The agreement is fairly good. In particular, the shock
position agrees remarkably well.

## 3.5 Direct iteration with shock

The direct iteration scheme, discussed in section 3.3 has been
extended recently to the case of profile flow with shock at zero
incidence (Niyogi and Das 1981). Converged solution with shock may
be obtained by the direct iteration scheme if the starting solution
contains a shock discontinuity or else if with a continuous starting
solution, the expansion shock which appears at the accelerating sonic
point be excluded at each iteration step. In the discontinuous starting solution the shock position is conveniently taken to be the
decelerating sonic point at the profile, delivered by the approximate

shock-free analytical solution Eq.(10), the strength being determined by the small perturbation form of the Rankine-Hugoniot condition in reduced coordinates

$$U_o^+ (X_S , 0) + U_o^- (X_S , 0) = 2,$$ (29)

where $X_S$ denotes the shock position and $U_o^-$ and $U_o^+$ denote respectively the starting values of U immediately before and after the shock. The shock determination in the first approach closely follows the procedure of Zierep (1962), discussed in section 2.4. The

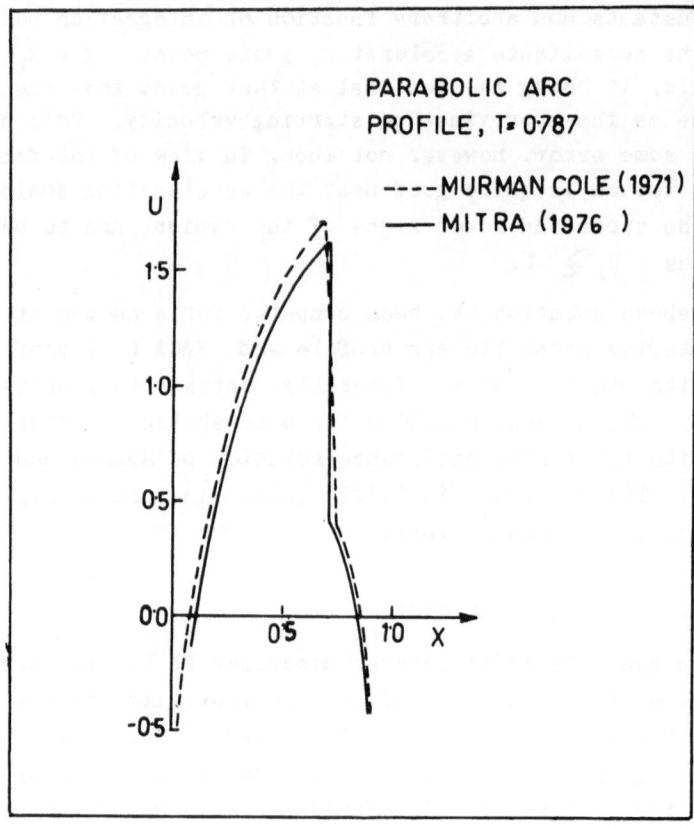

Fig.3.9  Reduced velocity distribution with shock for a parabolic arc profile T = 0.787. Comparison with finite difference solution of Murman and Cole (1971).

shock condition relevant to integral equation formulation, as established by Zierep is given by

$$2I_{n-1}(X_s) - \left[ 2U_p(X_s, 0) - 1 \right] = \frac{1}{2}\left[ (U_{n-1}^+)^2 + (U_{n-1}^-)^2 \right] - 1,$$

(30)

where $I_{n-1}$ denotes the value of the integral term in Eq.(18a) in
the $(n-1)$-th step, so that

$$I_{n-1}(X_s) = \int_0^1 \frac{U_{n-1}^2(\varepsilon, 0)}{2b(\varepsilon; U_{n-1}(\varepsilon, 0))} E\left[ \frac{\varepsilon - X_s}{b(\varepsilon; U_{n-1}(\varepsilon, 0))} \right] d\varepsilon,$$

(31)

Condition (30) is used along with the direct iteration scheme Eq.(18a),
and with these modifications, the direct iteration scheme converges
quite rapidly to a solution with shock and no expansion shock occurs.

An alternative approach for computing supercritical flows with
shock by direct iteration, is to start with the continuous starting
solution, Eq.(18b) and to exclude the expansion shock at each iteration
step. This can be achieved by prescribing a restriction at each iteration step on the growth of the velocity curve in the supercritical
region between the expansion shock at the accelerating sonic point and
the compression shock at the decelerating sonic point, determined according to the approximate solution (18b). The restrictions are to be
such that they should not impair the continuity of the slope and curvature of the velocity profile in the region between the expansion and
compression shocks. Thus at each iteration step, we correct the
computed U-values at the pivotal points lying between the expansion
and compression shocks. The corresponding algorithm may be described
as

$$\left[ (U_{n+1})_i \right]_{corr.} = \alpha_i (U_{n+1})_i, \quad 0 < \alpha_i \leq 1,$$

(32a)

the suffix 'corr' denoting corrected values. We now define

$$\beta = \left[ (U_{n+1})^-_{es} + \Delta^-_{es} \cdot 1 \right] \Big/ (U_{n+1})^+_{es} . \qquad (32b)$$

Here $\alpha_i$ are certain parameters, $(U_{n+1})^-_{es}$ and $(U_{n+1})^+_{es}$
denote the U-values at the $(n+1)$-th step before and after the
expansion shock, and $\Delta^-_{es}$ denote the gradient of U-curve at the
pivotal point immediately preceding the expansion shock and 1
denotes the interval length of the pivotal points. From computation,
we find that the following choice of parameters

$$\alpha_1 \le \beta , \qquad \text{and} \qquad \alpha_1 \geqslant \alpha_2 \geqslant \alpha_3 \geqslant \cdots \cdots \geqslant \alpha_m , \qquad (33c)$$

leads to converged supercritical flow with shock, where the suffix
m refers to the last pivotal point before the compression shock.

Both the alternative modifications of the direct iteration
scheme have been applied successfully in computing supercritical
transonic flow with shock past a parabolic arc profile and a NACA 0012
profile. The reduced velocity distribution along the profile chord
for a 6 per cent thick symmetrical parabolic arc profile at free-stream
Mach number $M_\infty = 0.87$ has been compared with the results of
Murman (1974) ( $M_\infty = 0.8715$) and the extended integral equation
solution of Nixon (1977) and the curved shock solution of Nixon (1979)
(to be discussed in Chapter IV) in Fig.3.10. The overall agreement of
the present solution, which converges in only fourteen iteration steps
requiring about 30 seconds of CPU time on a Burroughs B6700 computer
system, is better than the extended integral equation solution,
although the curved shock solution is better near the peak pressure.
Further, the present solution shows excellent agreement with the shock
position of Murman. Also, the expansion downstream of the shock is
better reproduced than that with the curved shock. From this result
and from our computational experience with other profile shapes, it
appears that the simplified model, Eq.(2.14), suggested by Oswatitsch
(1950) is sufficiently good for most practical purposes:

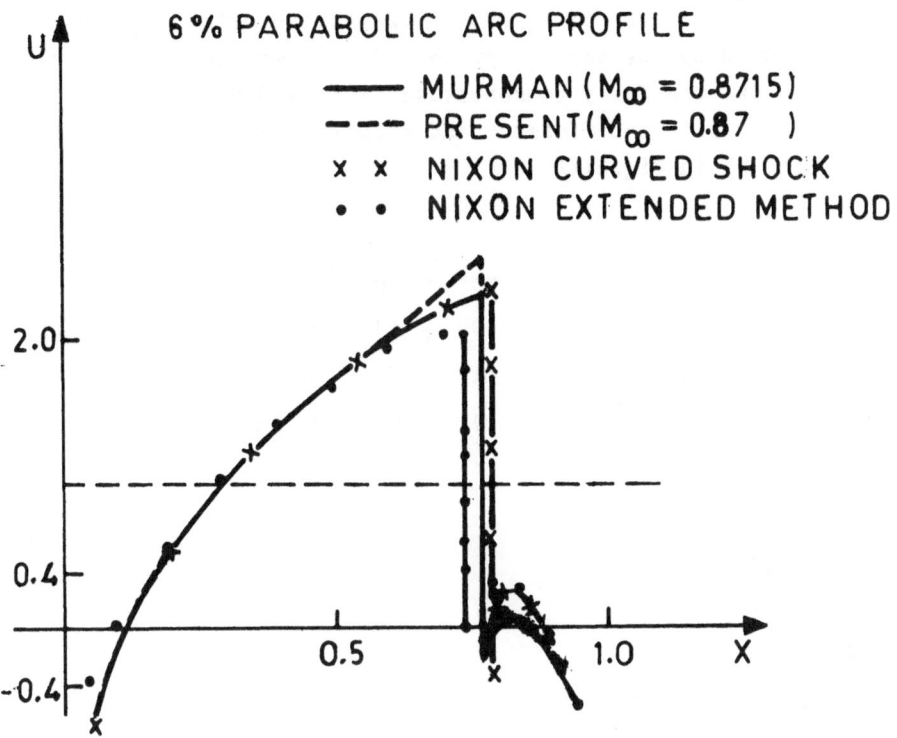

**Fig.3.10**   Reduced velocity distribution for a 6 per cent
parabolic arc profile with shock.  Comparison
with finite-difference solution of Murman
($M_\infty = 0.8715$), Nixon extended method and Nixon
curved shock method.

Choosing the parameter values

$$\alpha_1 = 0.95 = \beta ,$$

$$\alpha_i = \alpha_{i-1} - 0.05 , \quad i = 2, \ldots , m , \qquad (34)$$

the present solution for a NACA 0012 profile with $M_\infty = 0.8$ has been computed by the second approach discussed above. It has been compared with the fully conservative (FCR) and not fully conservative solutions by finite-difference and experimental results as given in Bailey (1975) in Fig.3.11.

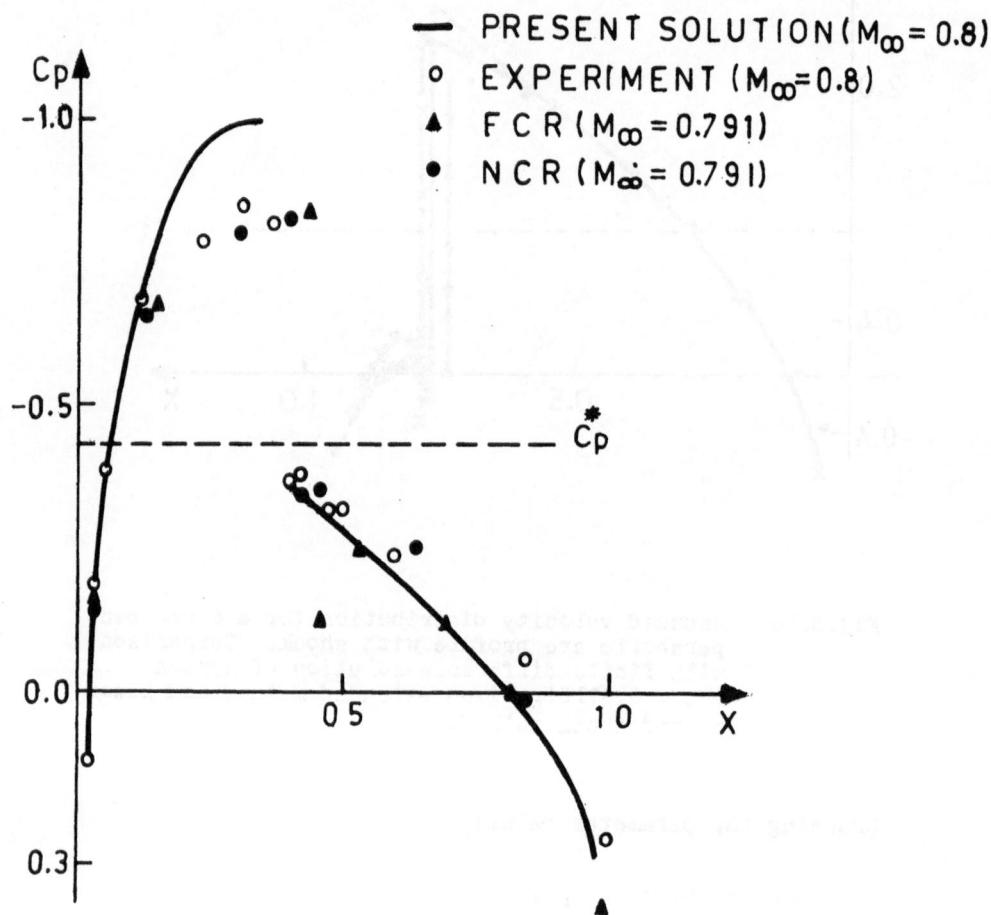

Fig.3.11  Reduced surface velocity distribution by direct iteration with shock for a NACA 0012 profile at $M_\infty = 0.8$. Comparison with finite-difference solution and experimental results (Bailey 1975).

The overall agreement of the present solution is satisfactory, although
the peak value before the shock indicates some error. However, the
shock position seems to agree very well with the experimental results.
It should be mentioned that for computing NACA 0012 profiles, a second
order edge correction as proposed by Nixon and Hancock (1974)

$$
U_p(X,\ 0)\Big]_{corr.} = \left[ \frac{1}{\mu} + U_p(X,\ 0) \right] \Bigg/ \left[ 1 + (\tau\mu\ \frac{dg}{dx})^2 \right]^{1/2} - \frac{1}{\mu}\ ,
$$

$$(35)$$

has been used.

The present solution indicates a great reduction in the number
of iteration steps which is a few hundred in the finite-difference
computations (Bailey 1975). In view of the simplified model used, the
present computations lead to an enormous reduction in CPU time roughly
by a factor of 60 compared to the finite difference procedure. For one
profile shape, Murman and Cole (1971) required about 30 minutes of CPU
time on an IBM 360 computer system, whereas the present computations
on Burroughs B6700 computer system which is almost equally fast as an
IBM 360 machine, require only about 30 seconds of CPU time. In view of
this, one can think of a hybrid direct iteration/finite-difference
procedure for results of high accuracy, which would use the converged
direct iteration solution of the simplified model as the starting
solution for the finite-difference procedure.

## 3.6 Method of Frohn

Frohn (1974) proposed an iterative method for solving the
integral equation of Oswatitsch, which introduces some new features for
computation of flows with shocks. The main idea of the method is to
break up the solution into a continuous part $U_c$ and a discontinuous
part $U_d$, as

$$
U = U_c + U_d\ .
$$

For this purpose, we consider the equation leading to the integral
equation of Oswatitsch

$$
U - I_{Sh}\left[ U \right] = U_p + \frac{1}{2}U^2 - I_{Sh}\left[ \frac{1}{2}U^2 \right] - I_G\ ,
$$

$$(36a)$$

where $I_G$ denotes the space integral

$$I_G = \frac{1}{2\pi} \int_{-\infty}^{\infty} \int_{-\infty}^{\infty} \frac{1}{2} U^2(\varepsilon, \eta) \frac{(\varepsilon - X)^2 - (\eta - Y)^2}{\left[(\varepsilon-X)^2+(\eta - Y)^2\right]^2} \, d\varepsilon \, d\eta .$$

(36b)

Further, $I_{Sh}\left[U\right]$ denotes the sum of the line integrals along the shock wave, arising out of application of Green's theorem to the small perturbation partial differential equation (1.21b). The term $I_{Sh}\left[\frac{1}{2} U^2\right]$ denotes the sum of the line integrals along the shock wave, which arise through integration by parts of the surface integral.

If shock conditions are applied we find that

$$I_{Sh}\left[U\right] = I_{Sh}\left[\frac{1}{2} U^2\right],$$

so that the shock integrals cancel out and the integral equation of Oswatitsch results. In the iterative solution of Frohn (1974, 1976) these line integrals are kept and the shock conditions are used as additional conditions, since the n-th step value of $I_{Sh}\left[\frac{1}{2} U^2\right]$ on the right of Eq.(36a) is not equal to the (n+1)-th step value of $I_{Sh}\left[U\right]$ appearing on the left. However, it is to be noted that when convergence occurs finally, they are equal.

It was observed by Frohn, that although the quantities $I_{Sh}\left[U\right]$ and $I_{Sh}\left[\frac{1}{2} U^2\right]$ are discontinuous across the shock, the function $U - I_{Sh}\left[U\right]$ is continuous and has a continuous first derivative across the shock. This function is denoted by $U_C$ and called the continuous part of the solution. The discontinuous function $I_{Sh}\left[U\right]$ is denoted by $U_d$ and called the discontinuous part of the solution. Eq.(36a) is solved iteratively, the double integral being approximated by a one-dimensional integral as in simplified Oswatitsch model.

Denoting by $U_{c1}$ a first approximation to $U_c$, the location of the shock wave, denoted by $S$, is assumed to be at the decelerating sonic point on the body determined by the first approximation $U_{c1}$. The first approximation to the discontinuous part at the shock wave is denoted by $U_{d1}(\pm S) = \pm D_1$, on the two sides of the shock. Then, since $U_{c1} = 1$ at the shock, it follows that

$$\frac{1}{2} - U_p(S) + \frac{1}{2} D_1^2 - I_G(S; U_{c1}, D_1) = 0. \tag{37}$$

This equation can be used for determining $D_1$. In some cases, it may happen that Eq.(37) has no real solution, indicating that the term $U_p(S)$ is too large. It can be shown (Frohn 1976) that the derivative of the left hand side of Eq.(37) with respect to the shock strength is zero. Thus, instead of solving Eq.(37), we may check for the minimum of $\frac{1}{2} D_1^2 - I_G(S; U_{c1}, D_1)$ and the term $U_p(S)$ may be eliminated. A second approximation of $U_c$ is then computed by

$$U_{c2} = U_p + \frac{1}{2} U_{c1}^2 + (U_{c1} - 1)U_{d1} + \frac{1}{2} U_{d1}^2 - I_G(S; U_{c1}, U_{d1}). \tag{38}$$

The whole procedure is then repeated iteratively.

Rapid convergence has been found by the application of this method. Only in a few cases convergence failed, presumably due to the non-existence of a solution of the approximated equation.

The first three iteration steps for a parabolic arc profile of the thickness ratio $\tau = 0.084$ at $M_\infty = 0.85$ have been presented in Fig.3.12. The computed solution particularly the shock position depends on how the double integral in Eq. (36a) is evaluated. Frohn (1974) gives an example of parabolic arc profile with shock, which indicate satisfactory agreement with finite-difference solution.

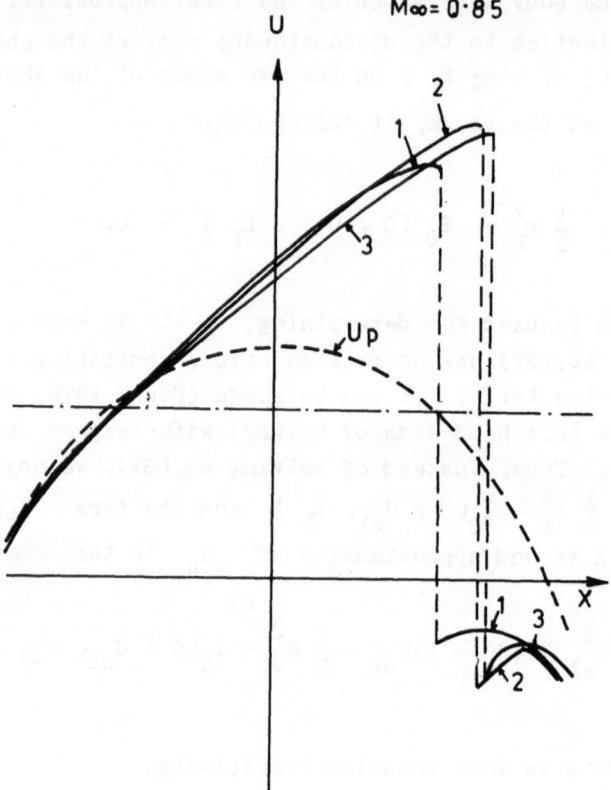

PARABOLIC ARC PROFILE
$\tau = 0.084$
$M_\infty = 0.85$

Fig.3.12 Surface velocity distribution of a parabolic arc profile in 3 iteration steps, after Frohn (1976)

## 3.7 Direct iteration of Oswatitsch equation

In view of the encouraging results obtained from the simplified model by means of the direct iteration scheme, and particularly in view of its satisfactory convergence behaviour we have started computations with the full equation (1) using the direct iteration scheme. As already pointed out, this equation is valid for shock-free flow as well as for flows with weak shocks, not necessarily straight and

normal, provided only that if a shock is present its curvature should
be small. For carrying out the computations, the flow plane is subdivided into a net work by straight lines parallel to the coordinate
axes. The double integral is evaluated numerically taking appropriate
care for the dipole singularity at the pivotal point $\mathcal{E} = X$, $\eta = Y$,
and the double integral is rewritten as

$$I(X,\ Y,\ \mathcal{E},\eta\ ;\ U) \equiv \frac{1}{2\pi} \int_{-\infty}^{\infty} \int_{-\infty}^{\infty} \frac{1}{2} U^2(\mathcal{E},\ \eta) K_2(\mathcal{E}-X,\ \eta-Y)\ d\mathcal{E}d\eta$$

$$\approx \frac{1}{2\pi} \int_{\eta=-B}^{B'} \int_{\mathcal{E}=-A}^{A'} \frac{1}{2}\left[ U^2(\mathcal{E},\ \eta) - U^2(X,Y)\right] K_2(\mathcal{E}-X,\ \eta-Y)\ d\mathcal{E}\ d\eta ,$$

(39)

where $A,\ A'$ and $B,\ B'$ are suitable finite values, such that the
contribution to the integral from regions outside it is negligible.
The direct iteration scheme is then defined as

$$U_{n+1}(X,\ Y) = U_p(X,\ Y) + \frac{1}{4} U_n^2\ (X,\ Y) - I(X,\ Y,\ \mathcal{E},\eta\ ;\ U_n),$$

$$n = 0,\ 1,\ 2,\ 3,\ \ldots\ ,$$

(40a)

with the starting solution

$$U_0(X,\ Y) = (\sqrt{3}+1)\left[1 - \left\{1 - (\sqrt{3}-1)U_p(X,\ Y)\right\}^{1/2}\right].$$

(40b)

The above scheme has been applied to a parabolic arc profile of
reduced thickness ratio $T = 0.67$, for which the solution is slightly
supercritical, the limits of integration along the X-axis being taken
as -1.5 and +2.5 and -1.5 and 1.5 in the Y-direction. The
domain of integration has been subdivided into 63 subintervals along
the X-axis and 60 along Y. The iteration converges correct up to
two decimal places in only 5 iteration steps. The reduced velocity
distribution on the profile axis presented in Fig.3.13, show that the
perturbations do not die down completely at $X = -1.5$ and $X = +2.5$,
and that somewhat greater values of the limits of integration $A,\ A'$
and $B,\ B'$ should be chosen. The critical value of the reduced thickness ratio is approximately 0.67 (slightly less than it). This

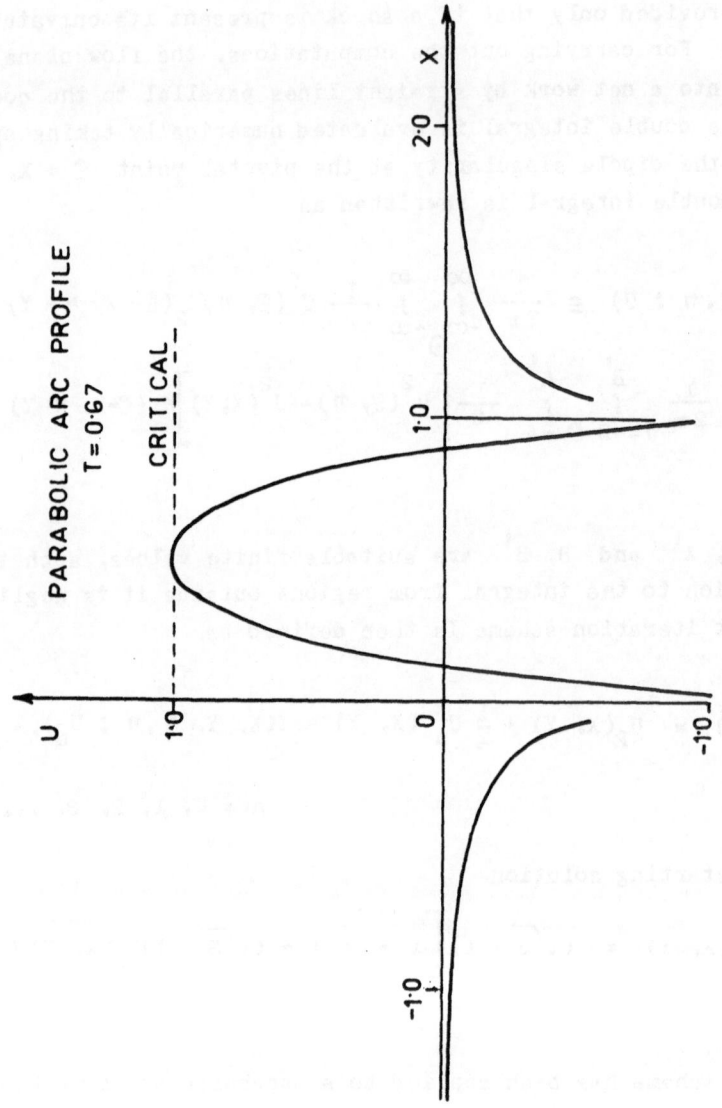

Fig.3.13    Reduced velocity distribution along the
axis of a parabolic arc profile computed
by direct iteration of Oswatitsch equation,
for  T = 0.67.

value is different from that delivered by the simplified Oswatitsch equation and is in favourable agreement with the value of Zierep(1962) which is slightly greater than 0.66. For higher values of the reduced thickness ratio also, the iteration converges quite rapidly. However, the converged solutions show an expansion shock and a compression shock symmetrically situated about the mid-chord of the profile. This point is being investigated at present.

Convergence of the direct iteration scheme Eqs.(40) has been studied analytically in section 7.2 in the space $L_2(E_2)$ by means of Banach Contraction Mapping Principle. It has been found that if the $L_2$-norm $\bar{U}_p$ of the known linearized solution does not exceed the value $\sqrt{2} - 1(\approx 0.414)$, the direct iteration scheme converges. Computations show that for a parabolic arc profile, the flow is supercritical for $\bar{U}_p > 0.32$, so that the above scheme converges for supercritical transonic flow.

# Chapter IV

## Unsymmetric Profiles at Small Incidence

### 4.1  Introduction

As already mentioned in section 1.7.1 there exists three
different forms of the integral equation formulation for transonic
flow past a thin unsymmetric profile at small incidence, two of which
are due to Nørstrud and the third due to Nixon and Hancock. It
should be noted that although Nixon and Hancock (1974) made the hypothesis of shock-free flow, the basic equations derived by them hold
also for weak shocks, not necessarily straight and normal, provided
only that the shock curvature be small, as shown in section 1.7.2.
Subsequently, Nixon (1975a), (1977) presented the 'extended method'
for effective treatment of shocks, which was modified further (Nixon
1979), to take into account the effect of small curvature of the
shock. It was noted in sections 1.7.1 - 1.7.5  that it is typical
of all these integral formulations for the lifting case, that the
velocity distribution along the upper and lower profile sides are
connected by means of a double integral to the unknown field velocity
distribution. For solving these equations, some knowledge of the
relationship between the field velocity distribution in terms of that
of the profile axis is needed. Such an information is not readily
provided by the basic equations, and simplifying assumptions are made
following the line of attack of Oswatitsch (1950). Various innova-
tions have been introduced in the Oswatitsch substitution Eq.(2.9)
and elaborate use have been made of electronic digital computers.
The present chapter is devoted to the methods of solution for the
unsymmetric lifting case and mainly the contributions of Nørstrud,
Nixon and Niyogi are outlined.

### 4.2  Method of Nørstrud

The integral formulations of Nørstrud for transonic lifting
profile flow have been presented in sections 1.7.3 and 1.7.5. The
basic equations of the first formulation are Eqs.(1.88) and Eqs.(1.89).
It is to be noted that for the lifting case, although  $\bar{U}^{+}$  is

identical with the symmetrical part of the Prandtl solution $U_p^+$, the antisymmetric part $\bar{U}^-$ is different from $U_p^-$. Consequently the unknowns in these equations are $U^+$, $U^-$ and $\bar{V}$, the antisymmetric part $\bar{U}_A$ being given by Eq.(89b) in terms of $\bar{U}^- = U^-$. This coupled system of nonlinear singular integral equations was solved by Nørstrud (1970) iteratively, by first approximating the double integrals by single integrals and then replacing the resulting single integrals numerically by a finite approximating sum. A system of nonlinear algebraic equations is obtained, which is solved by standard procedures like Newton's method. The velocity distribution at a field point was expressed in terms of that on the profile axis as

$$U(X, Y) = e^{-Y/r} \, U(X, 0+), \qquad 0 \leq Y < \infty, \qquad (1)$$

the parameter $r$ being determined by the irrotationality condition at the profile axis. The substitution (1) is very similar to that used by Oswatitsch (1950), viz. Eq. (2.9).

The second formulation of Nørstrud (1973b) is much simpler to treat mathematically and we present the method of Nørstrud as developed in this work. The basic integral equations of the problem are Eqs. (1.106). Substituting Eq.(1) in the double integral in Eqs.(1.106), the integration with respect to $\eta$ may be performed in closed form. We obtain from Eqs. (106)

$$U(X, 0\,\overset{+}{\underset{-}{}}) - \frac{1}{2} U^2(X, 0\,\overset{+}{\underset{-}{}}) = U_p(X, 0\,\overset{+}{\underset{-}{}}) \,\overset{+}{\underset{-}{}}\frac{1}{2}\left[\Delta U(X) - \Delta U_p(X)\right]$$

$$- \int_{\mathcal{E}=-\infty}^{\infty} U^2(\mathcal{E}, 0+) \, E(|\,X-\mathcal{E}\,|\,;\,r^+\,)d\mathcal{E}$$

$$- \int_{\mathcal{E}=-\infty}^{\infty} U^2(\mathcal{E}, 0-) \, E(|\,X-\mathcal{E}\,|\,;\,r^-\,)d\mathcal{E}, \qquad (2)$$

where the influence function $E(Z) = E(2|\,X - \mathcal{E}\,|\,/\,r)$ is defined as

$$E(|\,X-\mathcal{E}\,|;\,r) = \frac{1}{2\pi}\left\{ \sin(Z)\left[\frac{\pi}{2} - Si(Z)\right] - \cos(Z)\,Ci(Z)\right\},$$

$$(3)$$

the functions $Si$ and $Ci$ being integral sine and integral cosine functions.

The parameter $r^{\pm} = r(X, 0^{\pm})$ is determined by the irrotationality condition on the profile axis and the tangency boundary condition Eq.(1.24b) as

$$r(X, 0^{\pm}) = abs \left\{ U(X, 0^{\pm}) / \left[ F''(X) + G''(X) \right] \right\}. \tag{4}$$

The single integrals appearing in Eqs.(2) are integrated by quadrature, by approximating the range of integration with a finite number $N$ of discrete sub-intervals and representing the unknown function $U = U(X, 0^{\pm})$ by a mean value within each element. Neglecting influences from points upstream and downstream of the profile, Eqs. (2) may then be replaced approximately by the following system of non-linear algebraic equations:

$$U_i - \frac{1}{2} U_i^2 - \bar{U}_i + C_i + \sum_{j=1}^{2N} \varepsilon_{ij} U_j^2 = 0,$$

$$i = 1, 2, \ldots, 2N, \tag{5}$$

where we have used the symbol $\bar{U}$ in place of $U_p$. The influence coefficients $\varepsilon_{ij}$ are integral representations of the influence function $E(Z)$ and the function $C_i$ is a velocity correction function

$$C_i = \begin{cases} \frac{1}{2}(U_i - \bar{U}_i - U_{i+N} + \bar{U}_{i+N}), & i \leq N \\ \\ \frac{1}{2}(U_i - \bar{U}_i - U_{i-N} + \bar{U}_{i-N}), & i > N, \end{cases} \tag{6}$$

and the index $i \leq N$ identifies a point on the upper surface of the profile and the lower surface values are associated with the range $N < i \leq 2N$. The system of non-linear algebraic equations (5) was then solved numerically on an electronic digital computer. It is to be noted that in Nørstrud (1973b) there is a sign error in the term $C_i$ of Eq.(5), which has been corrected in Nørstrud (1973a).

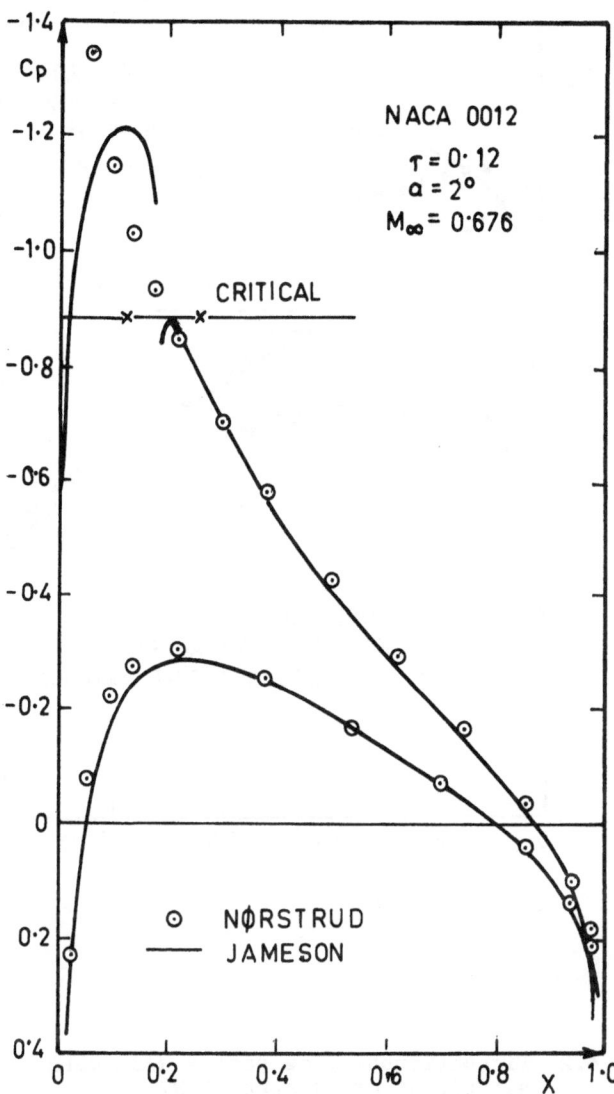

Fig.4.1   Supercritical lifting flow past a NACA0012
airfoil after Nørstrud (1973a). Comparison
with finite-difference solution of Jameson.

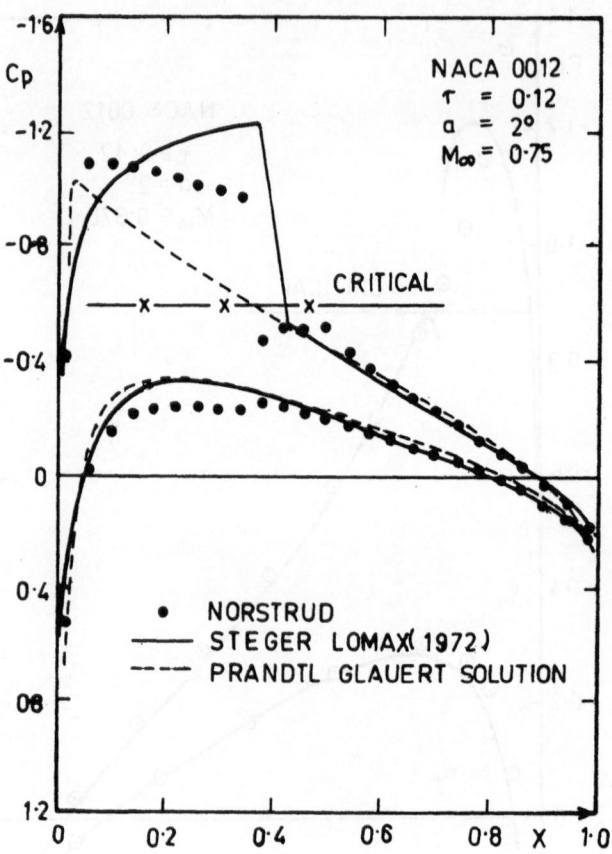

Fig.4.2   Discontinuous flow past a NACA 0012
          airfoil after Nørstrud (1973a).
          Comparison with finite-difference
          solution of Steger and Lomax (1972).

Solutions to the non-linear system (5) may be obtained by various iterative or relaxation procedures. For subcritical flow Newton's method is found to be the most convenient. However, application of Newton's method for supercritical flow is associated with the difficulty that the system (5) admits multivalued, discontinuous solutions. Since we are interested in obtaining discontinuous solutions which satisfy the entropy inequality, the introduction of an additional condition is necessary to make the solutions unique. This corresponds to the use of artificial viscosity in the various finite-difference procedures. For shock-free supercritical flow Nørstrud used the method of parametric differentiation (Yakovlev 1964, Rubbert and Landahl 1967, Schneider 1978) and the method of steepest descent for the case of supercritical flow with shocks.

Computational results have been presented in Fig.4.1 and 4.2 for a NACA 0012 profile at an incidence of $2^{\circ}$ , respectively for shock-free flow and for flow with shock.

It is to be seen from the above figures that the method of Nørstrud delivers fairly good results, although the lifting solution with shock is not very satisfactory. In this connection, it is to be mentioned that for lifting flow with shock, Nørstrud (1973a,b) introduced a correction term in the right hand side of the equations (1.65) and (1.106). However, from our calculations we see that Eqs.(1.65, 1.106) are valid for flows with shocks and no correction term seems necessary for describing lifting discontinuous flows.

### 4.3 Method of Nixon-Hancock

The basic integral equations of the formulation of Nixon-Hancock (1974) are Eqs. (1.75) and (1.76). For expressing the field velocity distribution in terms of that on the profile axis, they use the substitution

$$U(X, Y) = U(X, 0+) \Bigg/ \left[ 1 - \frac{Y\{f''(X) + g''(X)\}}{U(X, 0+)} \right], \quad Y \geqslant 0,$$

$$(7a)$$

and for the nonlifting case

$$U(X, Y) = U(X, 0+) \Bigg/ \left[ 1 - \frac{Y\{f''(X) + g''(X)\}}{2U(X, 0+)} \right]^2, \quad Y \geqslant 0,$$

$$(7b)$$

and similar substitutions for the lower part of the profile. With the help of these substitutions the double integrals in the basic equations are reduced to single integrals, following the path of Oswatitsch (1950). After a good amount of manipulation and simplification, and introducing edge correction, they ultimately derive the equation

$$U(X, 0 \pm) - \frac{1}{4} U^2(X, 0 \pm) = U_L(X, 0 \pm) + I(X, 0 \pm), \quad (8)$$

where $I(X, 0 \pm)$ is expressed in terms of the thickness integral $I_T$ and camber integral $I_C$ as

$$I(X, 0 \pm) = I_T(X, U(X, 0 \pm))$$

$$\pm \left[ \frac{1}{\pi} \left( \frac{1 - X}{X} \right)^{1/2} \int_0^1 \frac{I_C(\xi, U(X, 0 \pm))}{X - \xi} \left( \frac{\xi}{1 - \xi} \right)^{1/2} d\xi \right] /$$

$$\left[ 1 + \left\{ \frac{\beta^2 g'(X)}{\mu} \right\}^2 \right]^{1/2}, \quad \beta = (1 - M_\infty^2)^{1/2},$$

$$(9)$$

and $U_L$ denotes the linearized solution incorporating leading edge correction. For the expressions of the thickness and camber integrals $I_T$, $I_C$ and $U_L$ the reader is referred to the original work (Nixon and Hancock 1974).

For solving Eqs.(8), Nixon-Hancock start with the first approximation

$$U_0(X, 0 \pm) = 2 \left[ 1 - \left\{ 1 - U_L(X, 0\pm) \right\}^{1/2} \right], \quad (10a)$$

and calculate a second approximation

$$U_1(X, 0 \pm) = 2 \left[ 1 - \left\{ 1 - U_L(X, 0 \pm) - I_1(X, 0 \pm) \right\}^{1/2} \right],$$

$$(10b)$$

where $I_1(X, 0^{\pm})$ is obtained from $I$ by replacing $U$ by $U_0$ everywhere in Eq.(9). It should be noted that this iteration scheme is identical with the square-root iteration scheme defined by Eqs. (3.23) and (3.21a) already discussed in section 3.3 for the zero incidence simplified Oswatitsch equation. The leading edge correction is a new feature, and that it is applicable to lifting problem as well.

The above solution has been computed by Nixon and Patel (1975) for a number of profile shapes for shock-free supercritical flows, like NACA 0012 profile, NPL 3111, two Garabedian-Korn profiles and two NLR profiles. In most cases the second approximation gives results that are in fairly good agreement with exact results, especially in respect of the lift coefficient. The least accurate pressure distributions are for the rear loaded airfoils e.g. the NPL 3111 and the Garabedian-Korn, in which cases the second approximation failed to converge. Since only one iteration step in each case has been computed with this scheme, one cannot conclude that for the lifting supercritical flow the scheme converges. The method may be used with advantage for subcritical flow.

## 4.4 Extended method of Nixon

The method of Nixon-Hancock for shock-free flow was modified and extended by Nixon in a series of papers and the resulting procedure was named extended integral equation method (Nixon 1975a, 1976a, 1977, Nixon-Hancock, 1976). Second order terms were included in establishing the basic integral equations and the double integrals containing the field velocity distribution were evaluated more accurately, although neglecting the contributions of the velocity distributions in the semi-infinite regions before and after the profile, as in Oswatitsch substitution (2.9). Certain regularity conditions were established which ensure that the acceleration everywhere except across a shock wave is finite and continuous. This effectively excludes expansion shocks. In the process of satisfying these regularity conditions during the numerical solution of the integral equations, any shock wave that may appear in the flow field was assumed straight and normal to the free-stream. Following Nixon (1979), we present here the main features of the extended method for the simpler case of flow past a symmetric profile at zero incidence and refer to Nixon (1977) for the lifting case. The formulation containing second order terms may be found in Nixon and Hancock (1976).

The basic integral equation for a symmetric profile at zero incidence is expressed for the reduced velocity component $U(X, Y)$ as

$$U(X, Y) - \frac{1}{2} U^2(X, Y) = U_P(X, Y) + I_T(X, Y, X_S). \tag{11}$$

This equation is identical with the integral equation of Oswatitsch Eq. (1.78), where $X = X_S(Y)$ denotes the shock location, the pivotal point and the shock being excluded according to Oswatitsch principal value definition Eq. (1.54). If a shock is present, the correct shock location is determined by ensuring that there is a finite continuous acceleration everywhere except at the shock wave, a requirement which delivers the regularity conditions

$$\left[ U_P(X, Y) + I_T(X, Y, X_S) \right]_{X=X_0(Y)} = \frac{1}{2}, \tag{12a}$$

$$\frac{\partial}{\partial X} \left[ U_P(X, Y) + I_T(X, Y, X_S) \right]_{X=X_0(Y)} = 0, \tag{12b}$$

where $X = X_0(Y)$ denotes the unknown sonic line. Eqs. (11) and (12) completely determine the flow.

For numerical computation, the problem is slightly modified by the introduction of a parameter $\varepsilon(Y)$ such that Eqs. (11) and (12) are replaced by the equations

$$U(X, Y) - \frac{1}{2} U^2(X, Y) = U_P(X, Y) + \varepsilon(Y) I_T(X, Y, X_S), \tag{13}$$

and

$$\left[ U_P(X, Y) + \varepsilon(Y) I_T(X, Y, X_S) \right]_{X=X_0(Y)} = \frac{1}{2}, \tag{14a}$$

$$\frac{\partial}{\partial X} \left[ U_P(X, Y) + \varepsilon(Y) I_T(X, Y, X_S) \right]_{X=X_0(Y)} = 0. \tag{14b}$$

The double integral $I_T(X, Y, X_s)$ is evaluated by dividing the flow field into a finite number of strips parallel to the X-axis and approximating the distribution of the squares of the velocity in each strip in terms of the values on each strip edge by linear interpolation. The system of equations (13) and (14) then reduce to a set of algebraic equations for the velocity on the strip edges. The last strip is chosen to be situated at such a distance from the profile axis, that the contributions to the double integral from region outside is negligible. Further, the contributions from the semi-infinite regions before and after the profile are neglected.

The solution procedure as given in Nixon (1975) is as follows: First a normal shock location is assumed together with an initial guess for the field velocity. The field integral $I_T(X, Y, X_s)$ can then be evaluated and keeping the shock location constant, Eqs. (13), (14) are solved for new values of $U(X, Y)$, the parameter $\varepsilon(Y)$ being so chosen that Eqs.(14) are satisfied. This iterative procedure is repeated until the $\varepsilon(Y)$ are converged. The normal shock location is then moved until the $\varepsilon(Y)$ are close to unity (and $\frac{d\varepsilon}{dY} \approx 0$ ). Since a normal shock wave is assumed it is impossible to get $\varepsilon(Y)$ exactly unity (or $\frac{d\varepsilon}{dY} = 0$).

Results obtained by the use of the extended method indicate satisfactory agreement with more accurate results. Pressure distribution around a NACA 64A410 airfoil at zero incidence with free stream Mach number $M_\infty = 0.72$ computed by the extended method of Nixon (1977) has been presented in Fig.4.3, where the results of conservative and non-conservative finite-difference computations of Jameson are also presented. It is found that the results of Nixon agrees fairly with the incorrect non-conservative result rather than the correct conservative result. It is thought that this is due to the use of a much more simplified form of the potential equation than that used in the finite difference results. The computing time of the extended method is about half that of an equivalent relaxation finite-difference solution.

For a 6 per cent parabolic arc profile at $M_\infty = 0.8715$, the surface pressure distribution has been compared with the direct iteration solution of the simplified Oswatitsch equation by Niyogi and Das (1981) in Fig.3.10, where the curved shock solution of Nixon (1979) has

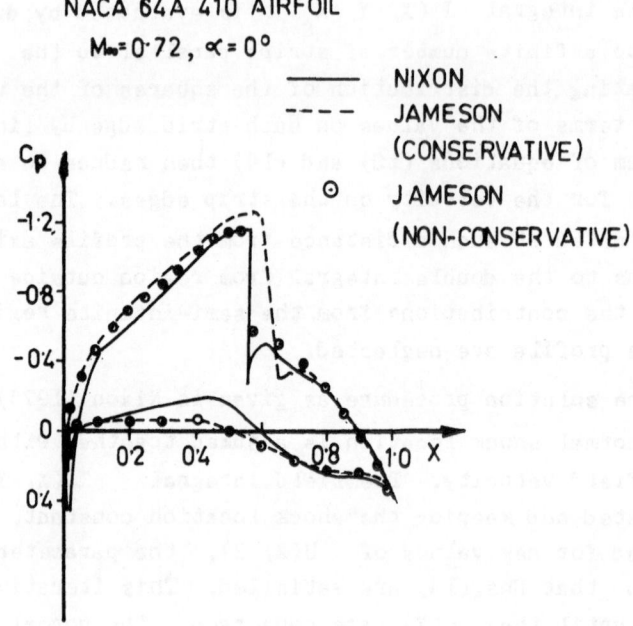

NACA 64A 410 AIRFOIL

$M_\infty = 0.72, \alpha = 0°$

——— NIXON

- - - - JAMESON
(CONSERVATIVE)

⊙ JAMESON
(NON-CONSERVATIVE)

Fig. 4.3  Pressure distribution about a NACA 64A410
airfoil at    $M_\infty$ = 0.72,    $\alpha$ = 0°,    after
extended method of Nixon (1977).  Comparison
with the results of Jameson.

also been presented for the sake of comparison.  It appears that the
great amount of computations associated with the extended method do
not bring in the expected accuracy in the computed solutions.  Particularly, the shock position indicates considerable error.  The errors
arise mainly from the neglect of the perturbation velocities in the
semi-infinite regions before and after the profile, which is also a
drawback of the simple Oswatitsch model.  In view of the rapid attenuation of the perturbation velocities away from the profile in the
Y-direction, the elaborate computations of the double integral in the

extended method of Nixon are hardly capable of achieving any mention-
able improvement over the Oswatitsch substitution. Further, in the
extended method the regularity conditions could not be satisfied
exactly, so that the integral equation solved is an approximate version
of the exact equation giving rise to a more serious source of errors
in determining the shock position. These errors have been essentially
removed in the curved shock solution which indicate excellent agreement
with the result of Murman (1974) apart from a small error immediately
behind the shock.

The curved shock solution of Nixon (1979) satisfies the regularity conditions on a curved shock with small curvature and evaluates
the field integral accordingly. Further, the effect of perturbation
velocities in the semi-infinite regions before and after the profile
has been included in evaluating the field integral. It is now possible
to satisfy the regularity conditions more accurately and excellent
result has been obtained for a parabolic arc profile. No computational
results for any other profiles have been presented in Nixon (1979).
It is also not mentioned if the CPU time needed compares favourably
with the finite-difference solutions. It would be interesting to see
if a lifting cambered airfoil with shock could also be computed with
such accuracy by the curved shock scheme at a lesser expenditure of
CPU time.

## 4.5  Approximate analytical solution for lifting flow with shock

For shock-free lifting flow, a simple extension of the zero
incidence approximate analytical solution Eq.(3.10) was given by
Chakraborty (1975) as

$$U_1^+(X, 0) = (\sqrt{3} + 1)\left[1 - \left\{ 1 - (\sqrt{3} - 1)\left[U_P^+(X, 0)\right.\right.\right.$$
$$\left.\left.\left. + U_A(X, 0)\right]\right\}^{1/2}\right]$$

(15a)

where

$$U_A(X, 0) = \frac{1}{2} \left\{ U_P^-(X, 0) \right\}^2$$

$$- \frac{1}{2\pi} \int_{-\infty}^{\infty} \int_{-\infty}^{\infty} \frac{1}{2} \left\{ U_P^-(\xi, \eta) \right\}^2 \frac{(\xi - X)^2 - \eta^2}{\left[ (\xi - X)^2 + \eta^2 \right]^2} \, d\xi \, d\eta,$$

(15b)

which is an approximate solution of the system of Eqs.(1.89) representing the first formulation of Nørstrud (1970). An assumption of the above solution is that the antisymmetric part of the linearized solution is the same as that of the Prandtl solution. For subcritical and lower supercritical Mach numbers at lower angles of incidence, this assumption does not introduce too much error, as may be seen from the various solutions computed by Chakraborty (1975). However, for higher angles of incidence, the error may not be negligible and an iteration would be necessary, where with advantage Eqs. (15) may be taken as starting solution.

The above shock-free solution has been extended recently, (Niyogi and Sen 1978) to the case of flow with shocks by adopting a procedure similar to that discussed in section 3.4. In place of Eq. (3.26), neglecting third and higher order small quantities in $t \sim 0 \, (\tau^{2/3})$, $\tau$ being the thickness ratio, we obtain

$$\frac{\partial}{\partial x} \left[ (\phi_{1x} - \mu) \, \bar{g}_x + \frac{1}{2} \bar{g}_x^2 \right] + \phi_{1x} \, \phi_{1xx} - \frac{\sqrt{3} - 1}{2} \, \phi_{1x}^+ \, \phi_{1xx}^+$$

$$- \mu^2 \frac{\partial}{\partial x}(U_P + U_A) - \mu \, \phi_{1xx}^- - \frac{\mu}{1 - M_\infty^2} \, \phi_{1yy} = 0,$$

(16)

where $\phi$ denotes the perturbation potential, $\phi_1$ the perturbation potential corresponding to the starting solution given by Eq.(15) and $\bar{g}(x, y)$ the correction term defined by

$$\emptyset(x, y) = \emptyset_1(x, y) + \bar{g}(x, y), \tag{17}$$

and $\mu$ is the Mach number function

$$\mu = \frac{1}{M_\infty^*} - 1.$$

Integrating and solving Eq.(16) as a quadratic equation for $\bar{g}_x$ , and following the line of approach described in the zero incidence case (section 3.4), we obtain finally in reduced coordinates, on the profile axis

$$U(X, 0) = 1 \pm \left[ \left\{ U_1(X, 0) - 1 \right\}^2 - \left\{ U_1^2(X, 0) - U_1^2(X_1^*, 0) \right\} \right.$$

$$+ \frac{\sqrt{3}-1}{2} \left( \left\{ U_1^+(X, 0) \right\}^2 - \left\{ U_1^+(X_1^*, 0) \right\}^2 \right)$$

$$+ 2 \left\{ \bar{U}_1^-(X, 0) - \bar{U}_1^-(X_1^*, 0) \right\}$$

$$+ 2 \left\{ U_P(X, 0) + U_A(X, 0) - U_P(X_1^*, 0) - U_A(X_1^*, 0) \right\}$$

$$+ 2 \int_{X_1^*}^{X} \frac{\partial V_1}{\partial Y} dX \right]^{1/2}, \tag{18}$$

where the upper and lower signs correspond to $U_1 \gtrless 1$ , and $X_1^*$ denote the abscissa of the accelerating sonic point on the body.

Figures 4.4 and 4.5 show the above solution for NACA 0012 profile for $M_\infty = 0.75$ at $2^\circ$ and $4^\circ$ angles of incidence and compared with the corresponding numerical solution of Garabedian and Korn (1971). The agreement is good. In both the figures, the shock position agrees particularly well, with that of Garabedian and Korn.

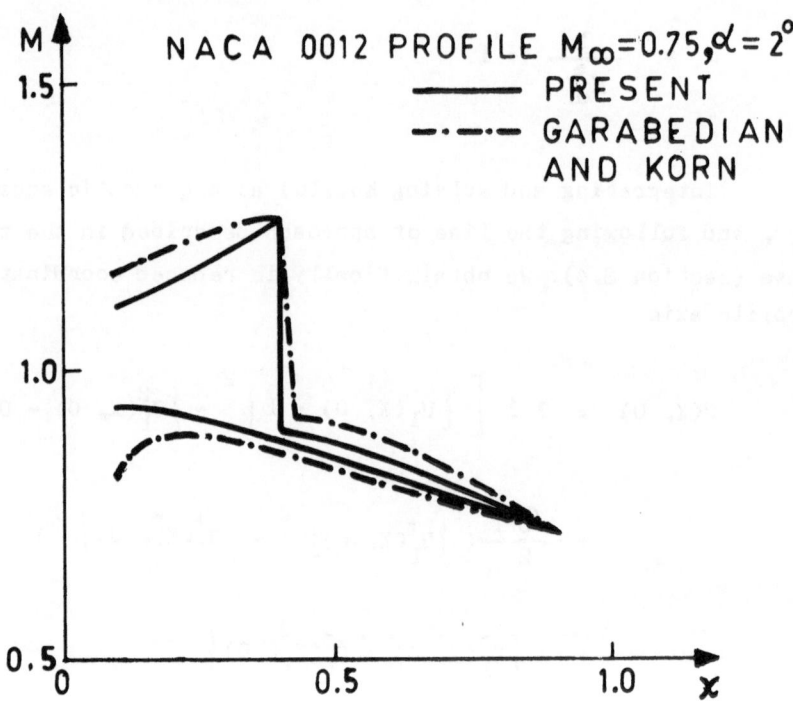

Fig.4.4 Surface pressure distribution for a NACA 0012 profile at $2^\circ$ incidence after Niyogi and Sen (1978). Comparison with solution of Garabedian and Korn (1971).

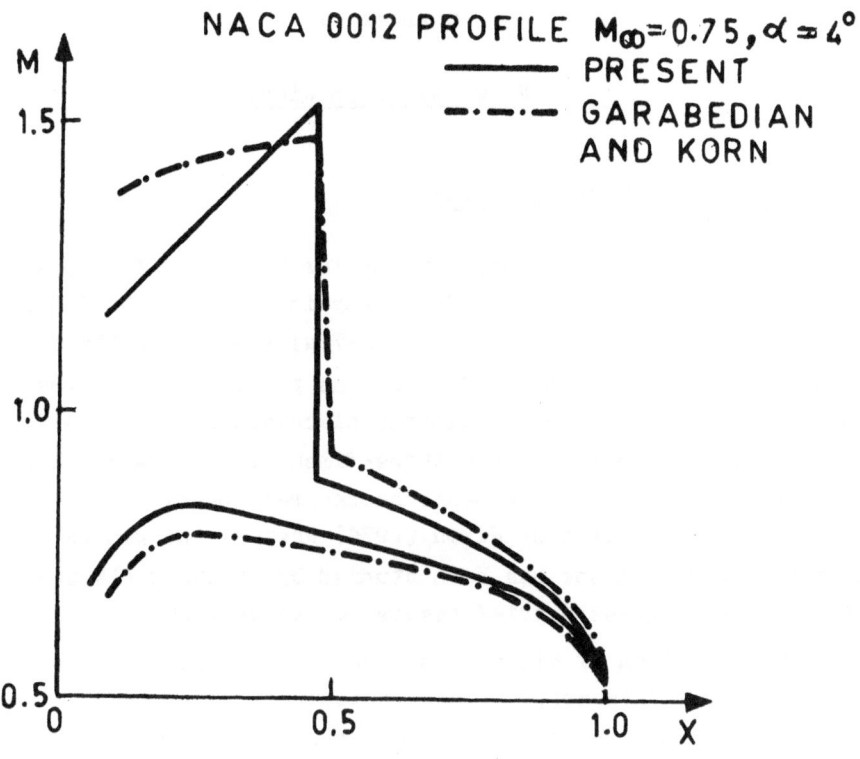

Fig.4.5   Surface pressure distribution of a
          NACA 0012 profile at   4°   incidence
          after Niyogi and Sen (1978).
          Comparison with solution of
          Garabedian and   Korn (1971).

However, the solution indicates poor agreement near the leading part
of the profile, which may be partly due to approximating the unsymmetric part of the linearized solution by the unsymmetric part of Prandtl
solution.  The present solution may be used as a good starting solution
for solving the coupled pair of two dimensional nonlinear singular
integral equations iteratively.

# Chapter V

Flow Past Thin Wings

## 5.1  Symmetrical Shock-free flow

Integral equation formulations for transonic flow past a thin finite wing have been presented already in section 1.6 of Chapter I, where the formulations of Nørstrud (1973a) and Nixon (1974) were given. The methods used for solving the problem of flow past thin profiles at zero and small incidence discussed in Chapters II - IV, may be readily extended to the three-dimensional case of flow past a thin finite wing. Zero incidence shock-free solution for a symmetric wing have been computed by Nixon (1974) whereas the general case of lifting flow with shock has been studied by Nørstrud (1973a). The present chapter gives a brief resume  of these works.

The solution of Nixon is a straight forward generalization of the corresponding two-dimensional procedure of Nixon and Hancock (1974). For reducing the volume integrals containing the unknown field velocity distribution  $U(X, Y, Z)$ , the substitution

$$\bar{g}(\varepsilon, \eta, \zeta) = \begin{cases} \bar{g}(\varepsilon, 0\pm, \zeta) \Big/ \left[ 1 + \dfrac{\eta}{b(\varepsilon, 0\pm, \zeta)} \right]^2, & \text{on } S_W, \\[3mm] 0, & \text{off } S_W, \end{cases} \tag{1}$$

was used, where  $\bar{g}(\varepsilon, \eta, \zeta)$  is a non-linear function of the velocity components containing  $\frac{1}{2} U^2$ ,  $V^2$  and  $W^2$ . The parameter b  is determined by the irrotationality condition on the wing planform $S_W$  as

$$b(\varepsilon, 0\pm, \eta) = \mp 2\,\bar{g}(\varepsilon, 0\pm, \eta) \Big/ \bar{g}_\eta(\varepsilon, 0\pm, \eta). \tag{2}$$

It is assumed that the distribution of $\bar{g}(\mathcal{E}, \eta, \varsigma)$ ahead of and behind the wing makes negligible contribution to the volume integral. Using the approximation Eqs. (1,2), the volume integrals in the basic equations (derived from Eq.(1.47) as indicated in section 1.6), may be reduced to surface integrals over the wing planform $S_W$. Subdividing the wing planform into trapezoidal panels, these integrals are approximated by finite sums. The final forms of the simplified equations are obtained as

$$U_T (X, Z) - \frac{1}{3} \bar{g}_T (X, Z) = U_P(X, Z) + I_{TX} (X, Z),$$

$$(3)$$

$$V_C(X, Z) - I_C(X, Z)$$

$$= -\frac{1}{4\pi} \iint_{S_W} \frac{U_C(\mathcal{E}, \varsigma)}{(Z - \varsigma)^2} \left\{ 1 + \frac{X - \mathcal{E}}{\left[(X - \mathcal{E})^2 + (Z-\varsigma)^2\right]^{1/2}} \right\} d\mathcal{E} \, d\varsigma,$$

$$(4)$$

$$W_T (X, Z) = W_P(X, Z) + I_{TZ}(X, Z),$$
$$(5)$$

where $U_P$ and $W_P$ are the corresponding linearized Prandtl solution for the velocity components, which are known quantities, and $U_T$, $V_C$, $\bar{g}_T$ and $W_T$ are defined as

$$U_T(X, Z) = \frac{1}{2}\left[U(X, 0+, Z) + U(X, 0-, Z)\right],$$

$$\bar{g}_T(X, Z) = \frac{1}{2}\left[\bar{g}(X, 0+, Z) + \bar{g}(X, 0-, Z)\right],$$

$$U_C(X, Z) = \frac{1}{2}\left[U(X, 0+, Z) - U(X, 0-, Z)\right],$$
$$(6)$$

$$V_C(X, Z) = \frac{1}{2}\left[V(X, 0+, Z) - V(X, 0-, Z)\right],$$

$$W_T(X, Z) = \frac{1}{2}\left[W(X, 0+, Z) + W(X, 0-, Z)\right]$$

the suffixes T and C being used to denote thickness and camber respectively. $I_{TX}$, $I_C$ and $I_{TZ}$ represent finite sums corresponding to thickness and camber integrals, the explicit forms of which are to be found in Nixon (1974).

Solutions have been computed by iteration for the nonlifting shock-free case only, when Eq.(4) drops out. The details of the iteration procedure is not available. The nonlinear function $\bar{g}$ has been taken as

$$\bar{g} = \frac{1}{2} U^2 + W^2 . \tag{7}$$

The results for a rectangular wing of aspect ratio 4 and $M_\infty = 0.75$, having a NACA 0012 section as computed by Nixon (1974) has been presented in Fig.5.1 and 5.2. The linear velocity components $U_P$ and $W_P$ are calculated from the linear potential equation with nominally exact boundary condition using a perturbation technique (Hewitt, private communication). It can be seen that the difference between the non-linear and linear pressure distributions is quite significant. The result obtained by neglecting the integral term in Eq.(1.57), as given by Eq.(10), are also shown in these figures as continuous curves. It may be seen that the simple solution Eq.(10) (computed by A.K. Roy) agrees very closely with the solution of Nixon.

It would be interesting to see how the results of Nixon (1974) compare with the finite difference solution or with that of Nørstrud (1973a). But no such comparison is available. Nixon (1974) has presented some comparison with the finite-difference solution for three subcritical cases of flow past RAE swept wings. These results are not presented here as they are not of much interest in the present context.

It is to be noted that neglecting the volume integral in the zero incidence symmetrical formulation (c.f. Eqs.(1.62) and (1.64b) )

$$U(X, Y, Z) = U_P(X, Y, Z) + \frac{1}{6} U^2(X, Y, Z)$$

$$- \int\limits_{-\infty}^{\infty} \int\limits_{-\infty}^{\infty} \int\limits_{-\infty}^{\infty} \frac{1}{2} U^2(\xi, \eta, \zeta) K_3(\xi - X, \eta - Y, \zeta - Z) d\xi \, d\eta \, d\zeta ,$$

$$\tag{8}$$

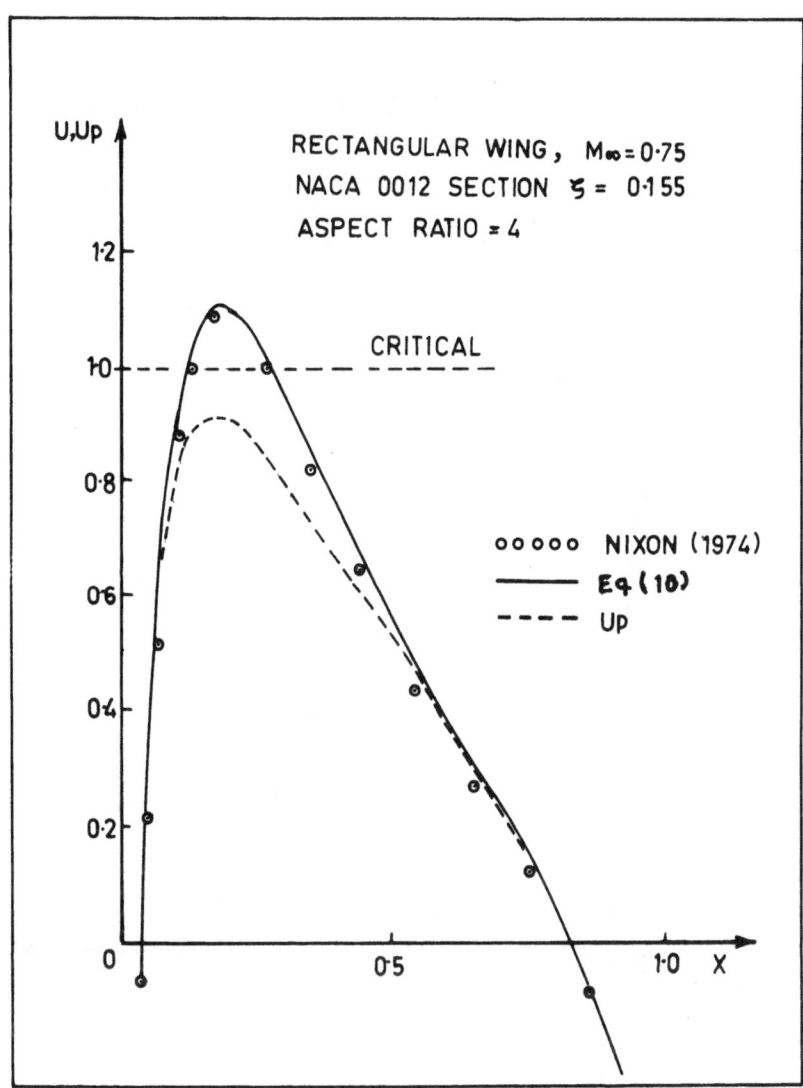

Fig.5.1 Pressure distribution on a rectangular
wing with NACA 0012 section of aspect
ratio 4 according to Eq.(10). Comparison
with the solution of Nixon (1974).
$M_\infty = 0.75$, $\alpha = 0^\circ$, at $\zeta = 0.155$

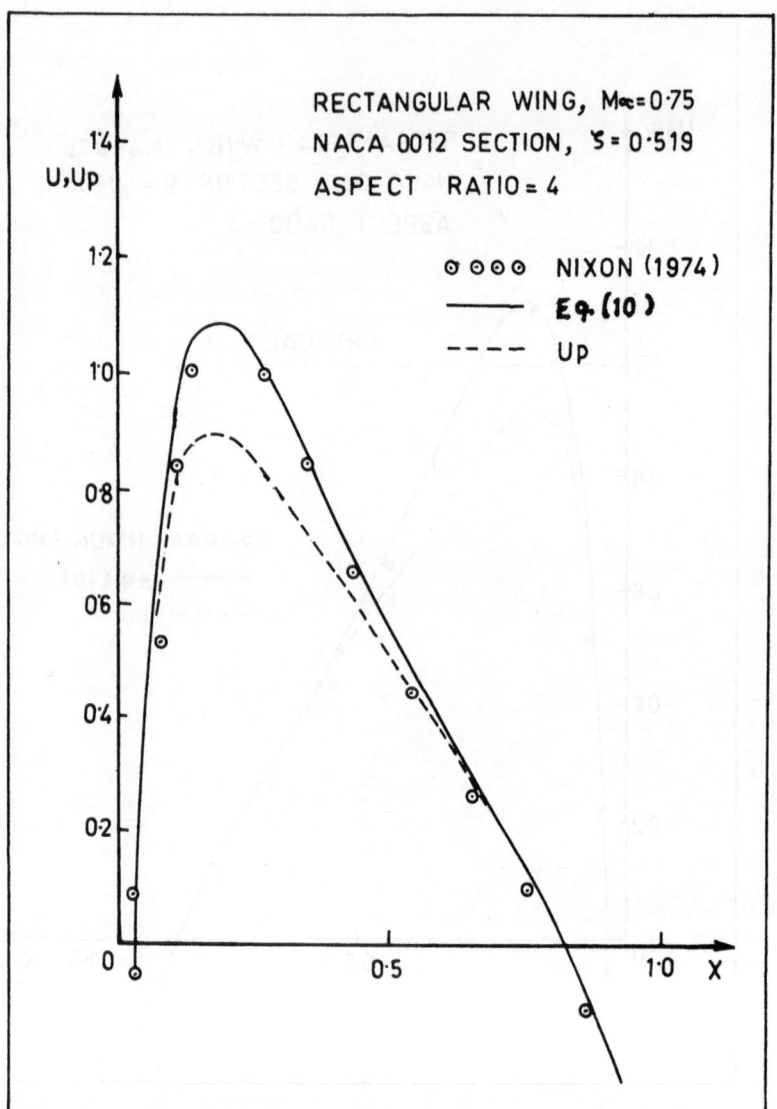

Fig.5.2 Pressure distribution on a rectangular wing
with NACA 0012 section, of aspect ratio 4
according to Eq.(10). Comparison with the
solution of Nixon (1974). $M_\infty = 0.75$,
$\alpha = 0^\circ$ at $\zeta = 0.519$

we get

$$U^2(X, Y, Z) - 6U(X, Y, Z) + 6\,U_p(X, Y, Z) = 0. \tag{9}$$

Solving it as a quadratic equation for $U(X, Y, Z)$, and neglecting the purely supersonic solution, it follows

$$U(X, Y, Z) = 3\left[1 - \left\{1 - \frac{2}{3}U_p(X, Y, Z)\right\}^{1/2}\right], \tag{10}$$

which is a simple approximate solution for subcritical and lower transonic shock-free flows. We have shown this solution in Fig.5.1 and 5.2, and compared with the solution of Nixon (1974). The agreement is good. Further, it is interesting to study the direct iteration scheme, which is generalized for the three-dimensional case as

$$U_{n+1}(X, Y, Z) = U_p(X, Y, Z) + \frac{1}{6}U_n^2(X, Y, Z)$$

$$- \int_{-\infty}^{\infty}\int_{-\infty}^{\infty}\int_{-\infty}^{\infty} \frac{1}{2}U_n^2(\xi,\eta,\zeta)K_3(\xi-X, \eta-Y, \zeta-Z)\,d\xi\,d\eta\,d\zeta,$$

$$n = 0, 1, 2, 3, \ldots, \tag{11}$$

where the starting solution is taken as that given by Eq.(10). Such a computation is at present under progress.

Convergence of the direct iteration scheme Eq.(11) has been studied analytically in section 7.3 in the space $L_2(E_3)$ by means of Banach Contraction Mapping Principle. It has been found that if the $L_2$-norm $\bar{U}_p$ of the known linearized solution does not exceed the value $3(\sqrt{2} - 1)$ the above scheme converges.

## 5.2 Lifting flow with shock

As already mentioned, Nørstrud (1973a) studied lifting flow past a thin unsymmetric wing at small incidence including shocks.

The basic equations in this formulation are given by Eqs. (1.65). In the particular case of flow past a thin symmetrical wing at zero incidence, we obtain only the Eq.(1.56). We emphasize here that Eqs.(1.65) are valid for lifting shock-free flow as well as for flows with weak shocks having small curvatures. For lifting flow with shock, Nørstrud (1973a) has included a correction term. The present author is unable to justify the inclusion of this correction term. On the other hand, the criticism of Nixon (1974) that the formulation of Nørstrud (1973a) for lifting flow is erroneous, is also not valid. The equations (1.65) determine completely the velocity distribution on the upper and lower sides of the wing for the zero incidence symmetric case as also for the lifting case. The appropriate tangency boundary condition and the trailing edge Kutta conditions are also included in these equations. Consequently, there is no error in the computations of Nørstrud (1973a) for the lifting shock-free case, apart from computational error and the error in approximating the field velocity distribution by means of that on the wing planform. However, such errors are also present in the works of Nixon, as discussed in the earlier chapters.

For expressing the field velocity distribution in terms of that on the wing planform, Nørstrud (1973a) used the substitution

$$U(X, \pm Y, Z) = U(X, \pm 0, Z) \exp \left\{ \mp Y \left/ r(X, \pm 0, Z) \right. \right\} ,$$
$$\tag{12}$$

where the parameter $r^{\pm}(X, Z) = r(X, \pm 0, Z)$ is approximately determined by the irrotationality condition and the tangency boundary condition Eq.(1.23) as

$$r^{\pm}(X, Z) = abs \left\{ (1 - M_{\infty}^2)^{3/2} U(X, \pm 0, Z) \left/ \left[ \bar{K} \frac{\partial^2}{\partial X^2} (Y_W^{\pm}) \right] \right. \right\} ,$$
$$\tag{13}$$

the wing shape being denoted by $Y_W^{\pm}$ (instead of $h_u$ and $h_l$ ).

Substituting it in Eqs. (1.65), the integration with respect to $\eta$ may be carried out in closed form. We obtain

$$U(X, 0, Z) - \frac{1}{2} U^2(X, 0, Z)$$

$$= \bar{U}(X, 0, Z) + \frac{1}{2} \left[ \Delta U - \Delta \bar{U} \right]$$

$$+ \frac{1}{4\pi} \int_{-\infty}^{\infty} \int_{-\infty}^{\infty} U^2(\mathcal{E}, +0, \varsigma) E^+ \, d\mathcal{E} \, d\varsigma$$

$$+ \frac{1}{4\pi} \int_{-\infty}^{\infty} \int_{-\infty}^{\infty} U^2(\mathcal{E}, -0, \varsigma) E^- \, d\mathcal{E} \, d\varsigma, \qquad (14)$$

where the kernel $E^{\pm} = E(|X - \mathcal{E}|, |Z - \varsigma|; r^{\pm})$ in the double integral reads after some simplification

$$E(|X - \mathcal{E}|, |Z - \varsigma|; r) = -\frac{2\pi}{r^2 \sigma} \left\{ \left[ \frac{4}{\sigma^2} \left( \frac{X - \mathcal{E}}{r} \right)^2 - 1 \right] \right. \cdot$$

$$\cdot \left[ \frac{2}{\pi} - H_1(\sigma) + N_1(\sigma) \right]$$

$$+ \frac{4}{\sigma} \left( \frac{X - \mathcal{E}}{r} \right)^2 \left[ H_0(\sigma) \right.$$

$$\left. - \frac{1}{\sigma}(H_1(\sigma) + N_1(\sigma)) + N_2(\sigma) \right] \Bigg\},$$

$$(15)$$

where

$$\sigma = 2 \left[ \left( \frac{X - \mathcal{E}}{r} \right)^2 + \left( \frac{Z - \varsigma}{r} \right)^2 \right]^{1/2}, \qquad (16)$$

and $N_\nu$ and $H_\nu$ are Neumann and Sturve functions of integer order $\nu$, whose values are available from standard tables (Roberts and Kaufmann 1966).

The solution of the non-linear integral equation (14) cannot be obtained analytically for arbitrary planform and thickness distribution, and numerical methods have to be used. For this, the upper and lower surfaces of the left (or right) half of the wing planform are subdivided into N number of trapezoidal panels as shown in Fig.5.3. The unknown function $U(X, \pm 0, Z)$ is assumed to be a constant within this specified panel boundaries.

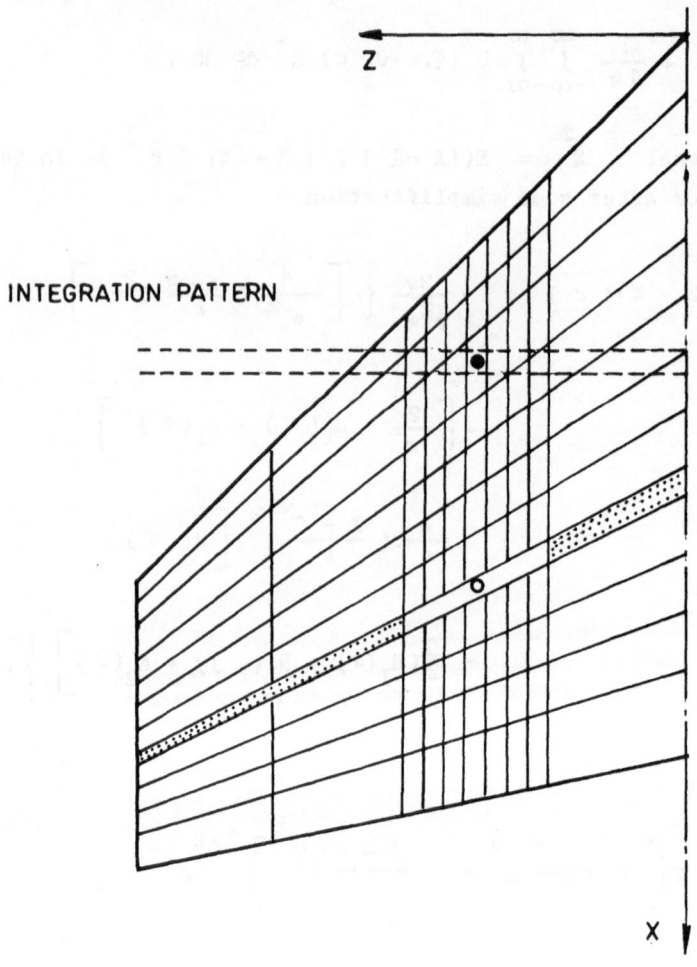

Fig.5.3  Subdivision of the wing planform into trapezoidal panels

Then the integral equation (14) is approximated by a system of $2N$ non-linear algebraic equations for the unknowns $U = U(X, 0, Z)$ at the pivotal points:

$$F_i(U_1, U_2, \ldots, U_{2N}) = U_i - \frac{1}{2} U_i^2 + C_i + \sum_{j=1}^{2N} \epsilon_{ij} U_j^2 = 0,$$

$$j = 1, 2, \ldots, 2N, \qquad (17)$$

where the index $i \leq N$ shall identify a value on the upper surface of the wing and the lower surface values are defined by the range $N < i \leq 2N$. The function $C_i$, $i = 1, 2, \ldots, 2N$ appearing in the system of Eqs. (17) is a velocity correction function associated with circulatory flows

$$C_i = \frac{1}{2} A_i, \qquad i = 1, 2, \ldots, 2N, \qquad (18a)$$

where

$$A_i = \begin{cases} U_i - \bar{U}_i - U_{i+N} + \bar{U}_{i+N}, & i \leq N, \\ \\ U_i - \bar{U}_i - U_{i-N} + \bar{U}_{i-N}, & i > N, \end{cases} \qquad (18b)$$

and $\epsilon_{ij}$ are the integral representations of the influence function $E = E(|X - \xi|, |Z - \zeta|; r)$, as evaluated over a surface element $\triangle S$.

As in the corresponding two dimensional case discussed in section 4.2, the system of equations (17) is solved by Newton's method for subcritical flow, by the method of parametric differentiation for continuous lifting flow and by the method of steepest descent in the case of lifting flow with shock. The details of these numerical methods may be found in Nørstrud (1973a). The linearized solution has been computed by the Woodward/Carmichael computer program (Hess 1969, Woodward 1968).

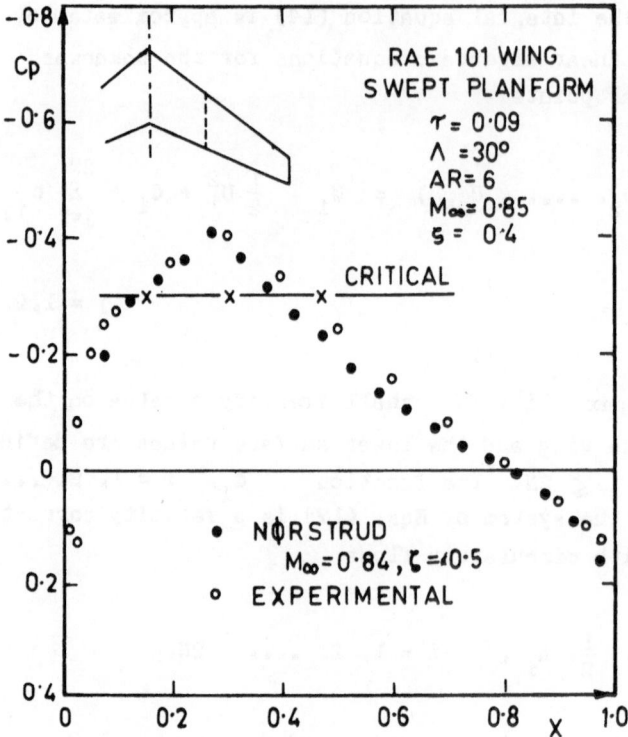

Fig.5.4    Continuous supercritical flow past a nonlifting
          swept wing after Nørstrud (1973a).  Comparison
          with experimental results.


        Computational results of Nørstrud (1973a) for a continuous
supercritical flow past a non-lifting  RAE 101 wing has been presented
in Fig.5.4, and compared with the experimental results of Jones (1971).
The agreement is very satisfactory.  Fig.5.5 shows the discontinuous
supercritical flow past a lifting rectangular wing of aspect ratio 4
having a parabolic profile as section.  These results are compared
with the finite-difference solution of Bailey and Steger (1972).  The
results are not so good as in the previous case.  The solution has
been computed using the method of steepest descent.  The slope of the
pressure curve indicates irregular behaviour near the mid-chord
region and it seems that a weak expansion shock has been formed near
the accelerating sonic line on the wing planform, which may be partly
responsible for the irregular slope.  Discontinuous flow past a nonlifting rectangular wing of aspect ratio 4 with parabolic arc section,

has been compared with the results of Bailey and Steger (1972) in
Fig.5.6. The agreement is fairly good apart from a small region near
the shock. The peak pressure before the shock indicates some discrepancy, which may be ascribed to the approximation of the field velocity in terms of that on the wing planform. However the shock
position indicates very good agreement. From these results, it may
be concluded that generally satisfactory results may be obtained from
the procedure of Nørstrud (1973a), although the treatment of the
lifting flow with shock needs modification. The information about
the computational time needed for a wing shape is not available. It

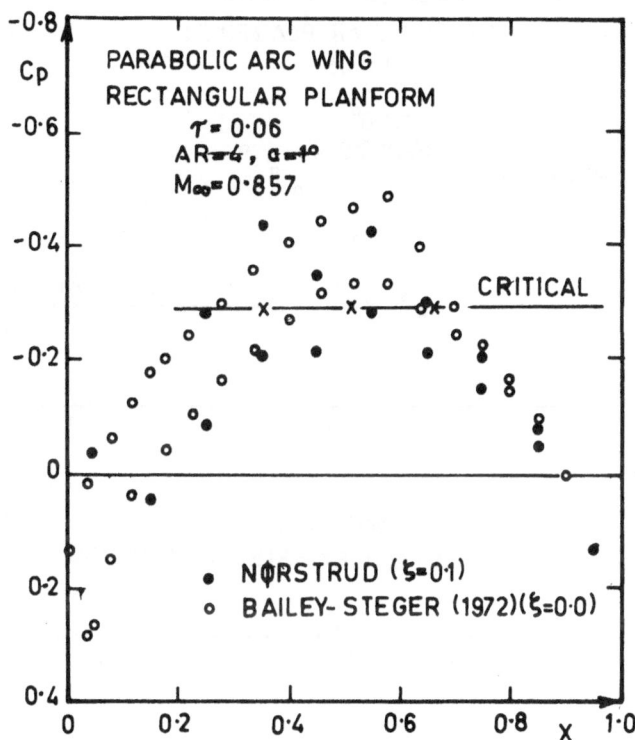

Fig.5.5   Supercritical flow past a lifting rectan-
          gular wing near the mid-span station
          after Nørstrud (1973a). Comparison with
          the finite-difference results of Bailey and
          Steger (1972).

would be interesting to know how the computational time compares
with that of the finite-difference procedure.

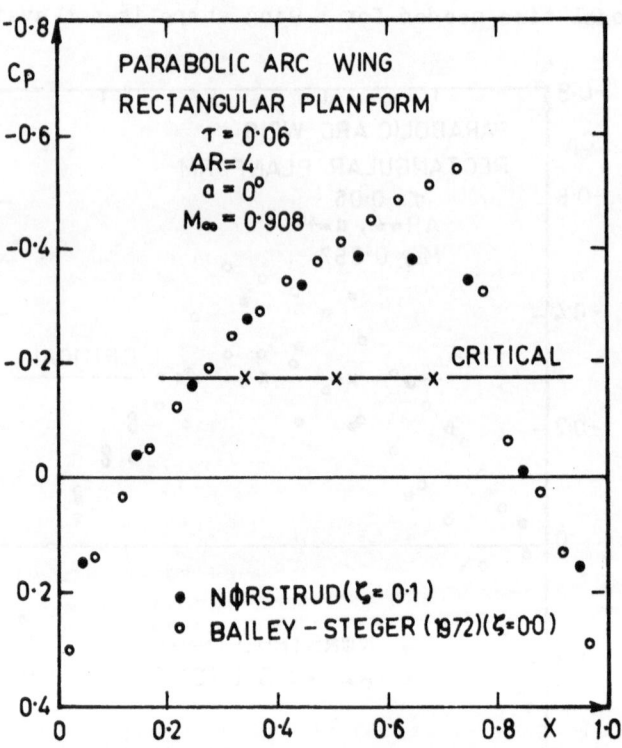

Fig.5.6   Discontinuous flow past a non-lifting
          rectangular wing near the mid-span
          station after Nørstrud (1973a).
          Comparison with the results of Bailey
          and Steger (1972).

# Chapter VI

Supercritical Airfoil Design

## 6.1  Symmetric shock-free airfoil design

Integral equation method has been used by Hansen (1976a,b) for the design of lifting airfoils with shocks. For symmetrical shock-free supercritical airfoil design a simple extension of the solution (3.10) has been given recently by Niyogi and Sen (1979, unpublished) which we discuss in the present section. The method of Hansen will be discussed in the next section.

The problem here is to determine the airfoil shape where the pressure distribution, or equivalently, the velocity distribution at the surface of the airfoil is prescribed. In the shock-free symmetrical case, Eq.(3.10) viz.

$$U(X, 0) = (\sqrt{3} + 1) \left[ 1 - \left\{ 1 - (\sqrt{3} - 1) U_p(X, 0) \right\}^{1/2} \right],$$

$$(1)$$

is known to be a fairly good approximation to the exact reduced velocity distribution. For the present problem, $U(X, 0)$ is prescribed and the profile shape is to be determined.

Solving Eq.(1) for the linearized Prandtl solution $U_p(X, 0)$, we obtain

$$U_p(X, 0) = U(X, 0) - \frac{\sqrt{3} - 1}{4} U^2(X, 0). \qquad (2)$$

Assuming the profile to be situated between $-1 \leq X \leq 1$ instead of $0 \leq X \leq 1$ used earlier, and putting $Y = 0$ in the integral representation Eq. (2.2) for $U_p(X, Y)$ and using the tangency boundary condition Eq.(2.3), it follows from Eq.(2)

$$\frac{1}{\pi} \int_{-1}^{1} \frac{\frac{dq(\xi)}{d\xi}}{X - \xi} d\xi = U(X, 0) - \frac{\sqrt{3} - 1}{4} U^2(X, 0), \qquad (3)$$

which is a Betz's integral equation (Oswatitsch 1956) for the unknown slope of the profile $q'(\Xi)$, for which closed form solution may be obtained. Solving it (Tricomi 1957) yields

$$T \frac{dq}{dX} = \frac{C}{\sqrt{1 - X^2}} + \frac{1}{\pi} \frac{1}{\sqrt{1 - X^2}} \int_{-1}^{1} U_P(t) \frac{(1 - t^2)^{1/2}}{t - X} dt,$$

(4)

where $U_P(t) \equiv U_P(t, 0)$ is given by Eq.(2) and $C$ is an arbitrary constant. Integrating Eq.(4) with respect to $X$ between -1 to 1, yields

$$T q(X) = C(\sin^{-1} X + \frac{\pi}{2}) + \frac{1}{\pi} \int_{-1}^{1} U_P(t) \sqrt{1 - t^2}$$

$$\cdot \left[ \int_{-1}^{X} \frac{ds}{(t - s) \sqrt{1 - s^2}} \right] dt.$$

(5)

Using the condition that the profile should be closed at the trailing edge

$$q(X) = 0, \quad \text{at} \quad X = 1,$$

(6)

the arbitrary constant $C$ in Eq.(4) vanishes. Performing the integration with respect to $s$ in Eq.(5), we finally obtain for the profile shape

$$T q(X) = \frac{1}{\pi} \int_{-1}^{1} U_P(t) \ln \left| \frac{\sqrt{(1 - X)(1 + t)} + \sqrt{(1 - t)(1 + X)}}{\sqrt{(1 - X)(1 + t)} - \sqrt{(1 - t)(1 + X)}} \right| dt,$$

(7)

which is reduced to a simple integration. In view of the relation between reduced and unreduced thickness ratio

$$T = \tau \left/ \left[ (1 - M_\infty^2)(\frac{1}{M_\infty^*} - 1) \right] \right. ,$$

(8)

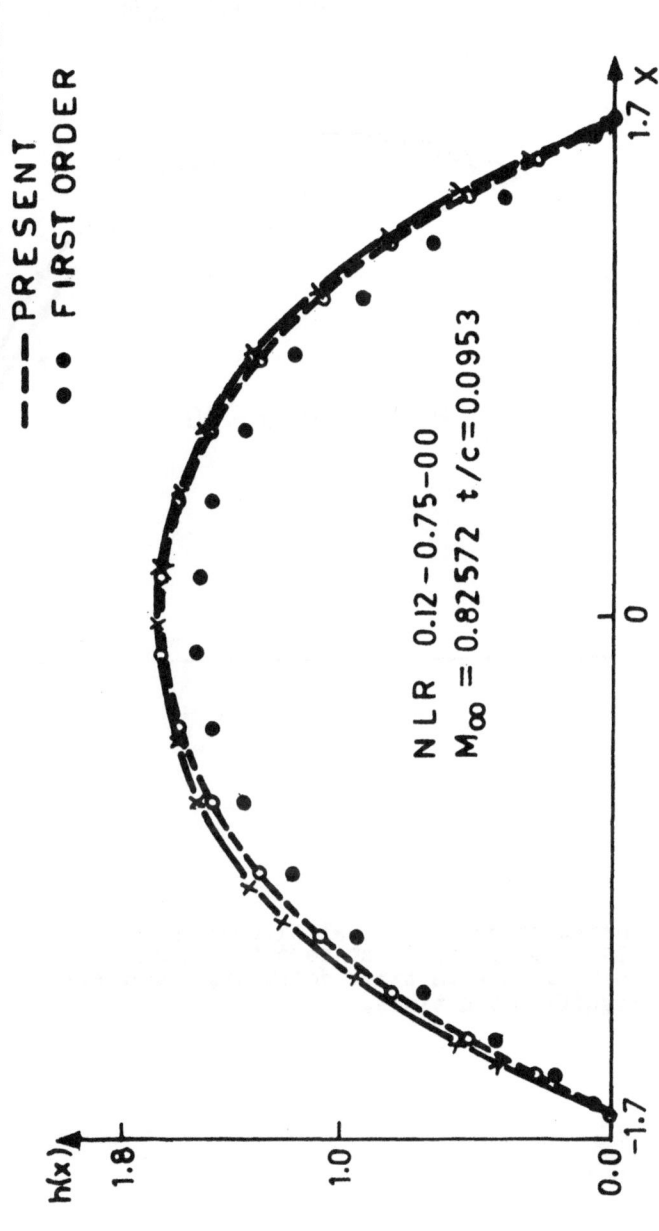

Fig.6.1 Design of supercritical Nieuwland profile NLR 0.12 - 0.75 - 00 at zero
incidence. Comparison with he exact solution. Ordinates magnified ten times.

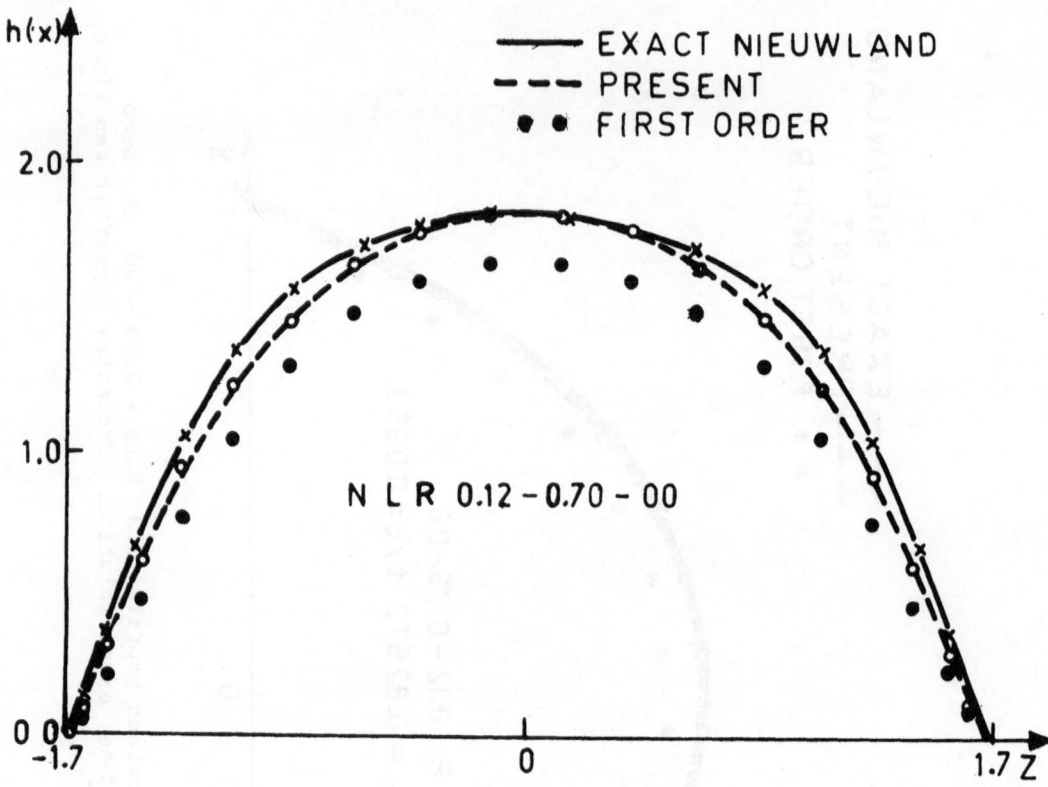

Fig.6.2  Design of supercritical Nieuwland profile
NLR 0.12-0. 70.00  at zero indicence.
Comparison with exact solution.  Ordinates
magnified ten times.

the actual profile shape may be obtained from Eq.(7), where the value
of $U_p(t)$ is to be taken from Eq.(2). Computations carried out
using Eq.(7) with twenty pivotal points on the profile axis show that
it is of moderate accuracy. This may be improved by performing an
iteration step with the simplified Oswatitsch model (discussed in
section 2.2):

$$U_p(X) = U_0(X) - \frac{1}{2} U_0^2(X) + \int_{-1}^{1} \frac{U_0^2(\mathcal{E})}{2\, b(\mathcal{E})} \, E(\frac{\mathcal{E} - X}{b(\mathcal{E})}) \, d\mathcal{E}.$$

$$(9)$$

The value of the parameter b may be determined from the first order
profile shape Eq.(7), using Eq.(2.11a). Results of adequate accuracy
may be obtained, if following the idea of Oswatitsch (1950), the value
of the parameter b is taken as a constant, equal to the maximum value
of $U_p$ determined by Eq.(2). The value of $U_p$ so determined from
Eq.(9) may be used in Eq.(7), to determine the improved value of the
profile shape.

The above procedure has been applied by S.Basu (Basu and Niyogi
1980) to a number of profile shapes including Nieuwland profiles.
Generally satisfactory results have been obtained. For Nieuwland
profiles, the velocity distribution has been approximated by Cubic
Splines and the integrations in Eqs. (7, 9) have been performed by
20-point Gaussian quadrature formula, paying due attention to the
singularity. The results for two such Nieuwland profiles are presented
in Fig.6.1 and Fig.6.2, where the exact solution of Nieuwland (1967)
as computed by Boerstoel (1968) and the first order solution Eq.(7)
are also shown. Both the cases show excellent agreement.

## 6.2  Method of Hansen

For the general case of lifting airfoils with shock, Hansen
(1976a,b) has developed a method for airfoil design based on the
integral formulation. He starts with a modified version of the small
perturbation equations, and arrives at a somewhat different form of
the basic integral equation. The essential features of Hansen's
method are discussed in this section.

For steady plane inviscid flow, the continuity equation (1.12c)
may be approximated as

$$\left(1 - \frac{u^2}{c^2}\right)\frac{\partial u}{\partial x} + \frac{\partial v}{\partial y} = 0, \tag{10}$$

on the assumption that the velocity component $v$ is negligibly small compared to the sound speed $c$. Using Bernoulli's equation (1.8), and the irrotationality condition (1.12d), Eq.(10) may be expressed in terms of the perturbation potential $\emptyset(x, y)$ as

$$\frac{\partial^2 \emptyset}{\partial x^2} + \frac{\partial^2 \emptyset}{\partial y^2} = \frac{2}{(\gamma-1)\beta^2}\left[\frac{(1 + \frac{\partial \emptyset}{\partial x})^2}{m^2-(1 + \frac{\partial \emptyset}{\partial x})^2} - \frac{1}{m^2 - 1}\right]\frac{\partial^2 \emptyset}{\partial x^2}, \tag{11}$$

where the parameters $m$ and $\beta$ are defined as

$$m^2 = 1 + \frac{2}{(\gamma-1) M_\infty^2}, \qquad \beta = (1 - M_\infty^2)^{1/2}, \tag{12a}$$

and the function $\rho$ is defined by

$$\frac{\partial \emptyset}{\partial x} = \frac{u - u_\infty}{u_\infty}, \qquad \frac{\partial \rho}{\partial y} = \frac{v}{u_\infty \beta}. \tag{12b}$$

Further the y-coordinate is stretched by a factor $\beta$ and the same symbols are used for the new coordinates. Eq.(11) is valid in the transonic range for small perturbation as also in a neighbourhood of the stagnation point when it reduces to Laplaces equation. Thus it describes reasonably well the supercritical transonic flow past a thin profile with blunt leading edge.

Applying Green's theorem Eq.(1.31) to the functions $\emptyset$ and $\ln\left\{(x - \varepsilon)^2 + (y - \eta)^2\right\}^{1/2}$ and following a procedure similar to that adopted in section 1.5.1 for the three-dimensional case, Eq.(11) is converted to the integral equation

$$\emptyset(x,y) = \frac{1}{2\pi}\left[\int_0^1 \left\{\left(\frac{\partial\emptyset}{\partial\eta}\right)_u - \left(\frac{\partial\emptyset}{\partial\eta}\right)_1\right\} \ln\sqrt{(\xi-x)^2+y^2}\ d\xi\right.$$

$$\left. - \int_0^1 \left\{\left(\frac{\partial\emptyset}{\partial\xi}\right)_u - \left(\frac{\partial\emptyset}{\partial\xi}\right)_1\right\} \tan^{-1}\frac{\xi-x}{y}\ d\xi\right]$$

$$+ \frac{1}{\pi(\gamma-1)\beta^2}\int_{-\infty}^{\infty}\int_{-\infty}^{\infty}\left\{\frac{\left(1+\frac{\partial\emptyset}{\partial\xi}\right)^2}{m^2-\left(1+\frac{\partial\emptyset}{\partial\xi}\right)^2} - \frac{1}{m^2-1}\right\}\cdot$$

$$\cdot\frac{\partial^2\emptyset}{\partial\xi^2}\ln\sqrt{(\xi-x)^2+(\eta-y)^2}\ d\xi\ d\eta - \frac{1}{4}\Gamma, \qquad (13)$$

where the suffixes $u$ and $1$ refer to the quantities on the upper and lower profile sides, and $\Gamma$ represents the dimensionless total circulation.

The first term within the square bracket in Eq.(13), denoted by $\emptyset_H$, represents the solution of the homogeneous equation, i.e. Laplaces equation. The second term involving the double integral is denoted by $\emptyset_N(x, y)$. It represents the potential at the point $P(x,y)$ induced by a source distribution over the entire flow field, the local intensity of the source being proportional to the non-linear term of equation (11). It is a non-linear function of the local velocity. The perturbation velocities induced by the source distribution over the field is continuous over the entire field. In particular, the induced perturbation velocities on the upper and lower profile sides are equal to each other:

$$\left(\frac{\partial\emptyset_N}{\partial x}\right)_u = \left(\frac{\partial\emptyset_N}{\partial x}\right)_1, \qquad \left(\frac{\partial\emptyset_N}{\partial y}\right)_u = \left(\frac{\partial\emptyset_N}{\partial y}\right)_1. \qquad (14)$$

Consequently, the non-linear source distribution over the field do not contribute to the jump in the tangential and normal velocities,

$$\left(\frac{\partial\emptyset}{\partial\eta}\right)_u - \left(\frac{\partial\emptyset}{\partial\eta}\right)_1 \quad\text{and}\quad \left(\frac{\partial\emptyset}{\partial\xi}\right)_u - \left(\frac{\partial\emptyset}{\partial\xi}\right)_1,$$

in the integrand of Eq.(13). Thus the first term $\emptyset_H(x, y)$ in Eq.(13) is independent of $\emptyset_N(x, y)$ and as such the potential at a point $P(x, y)$ of a non-linear compressible flow past a profile, described by Eq.(11) may be represented as a superposition of the potentials $\emptyset_H$ and $\emptyset_N$ :

$$\emptyset(x, y) = \emptyset_H(x, y) + \emptyset_N(x, y). \tag{15}$$

This idea simplifies the design problem in the lifting case.

It appears convenient to work with the velocity components, instead of the potential, since for the airfoil design problem, the velocity distribution on the profile contour $y_p = 0$ is prescribed. Differentiating Eq.(13) with respect to $x$ and $y$, and carrying out an integration by parts, it follows for $y = 0$ (correcting a misprint in Hansen (1976a, b) ):

$$\bar{u}(x, 0) = \bar{u}_H(x, 0) + \bar{u}_N(x, 0)$$

$$= \bar{u}_H(x, 0) + \frac{2}{(\gamma-1)\beta^2}\left[-\frac{m^2}{m^2-1}\bar{u}(x, 0) + \frac{m}{2}(\ln\frac{m+1+\bar{u}(x,0)}{m-1-\bar{u}(x,0)} - \ln\frac{m+1}{m-1})\right]$$

$$- \frac{1}{\pi(\gamma-1)\beta^2} \int_{-\infty}^{\infty}\int_{-\infty}^{\infty}\left[-\frac{m^2}{m^2-1}\bar{u}(\xi, \eta) + \frac{m}{2}(\ln\frac{m+1+\bar{u}(\xi,\eta)}{m-1-\bar{u}(\xi,\eta)} - \ln\frac{m+1}{m-1})\right]$$

$$\cdot \frac{(\xi-x)^2 - \eta^2}{\left[(\xi-x)^2 + \eta^2\right]^2} d\xi\, d\eta, \tag{16}$$

$$\bar{v}(x,0) = \bar{v}_H(x,0) + \bar{v}_N(x,0)$$

$$= \bar{v}_H(x,0) - \frac{2}{\pi(\gamma-1)\beta^2}\int_{-\infty}^{\infty}\int_{-\infty}^{\infty}\left[-\frac{m^2}{m^2-1}\bar{u}(\xi, \eta)\right.$$

$$\left. + \frac{m}{2}(\ln\frac{m+1+\bar{u}(\xi,\eta)}{m-1-\bar{u}(\xi,\eta)} - \ln\frac{m+1}{m-1})\right] \cdot \frac{(\xi-x)^2 - \eta^2}{\left[(\xi-x)^2 + \eta^2\right]^2} d\xi\, d\eta,$$

$$\tag{17}$$

where

$$\bar{u} = \frac{u - u_\infty}{u_\infty}, \qquad \bar{v} = \frac{v}{u_\infty \beta} . \qquad (18)$$

The unknown field velocity distribution $\bar{u}(\xi, \eta)$ appears under the integral sign in Eqs. (16, 17). This is expressed in terms of that on the profile axis by the approximate relations

$$\bar{u}(x, y) = 0, \qquad \text{for} \quad -\infty \leq y \leq -b(x),$$

$$\bar{u}(x, y) = \bar{u}(x, 0)(1 - \frac{y}{b(x)}), \qquad \text{for} \quad -b(x) \leq y \leq 0,$$

$$\bar{u}(x, y) = \bar{u}(x, 0)(1 + \frac{y}{b(x)}), \qquad \text{for} \quad 0 \leq y \leq b(x),$$

$$\bar{u}(x, y) = 0, \qquad \text{for} \quad b(x) \leq y \leq \infty, \qquad (19)$$

the parameter $b(x)$ being determined by the irrotationality condition, and the exact tangency boundary condition.

With the help of the substitutions Eq.(19), the integration with respect to $\eta$ in Eqs.(16, 17) may be carried out in closed form. It follows then from Eqs. (16, 17)

$$\bar{u}_H(x, 0) = \bar{u}(x,0) - \bar{u}_N(x,0)$$

$$= \bar{u}(x,0) - f_1(\bar{u}(x,0)) - \left[ \int_0^1 f_2(\xi,x)d\xi \right]_u - \left[ \int_0^1 f_2(\xi,x)d\xi \right]_1 ,$$

$$(20)$$

$$\bar{v}_H(x, 0) = \bar{v}(x,0) - \bar{v}_N(x,0)$$

$$= \bar{v}(x,0) - \left[ \int_0^1 f_3(\xi,x) d\xi \right]_u - \left[ \int_0^1 f_3(\xi,x) d\xi \right]_1 . \qquad (21)$$

The explicit forms of the functions $f_1$, $f_2$ and $f_3$ may be found in Hansen(1975). Eqs.(20,21) are the basic equations for the design method of Hansen.

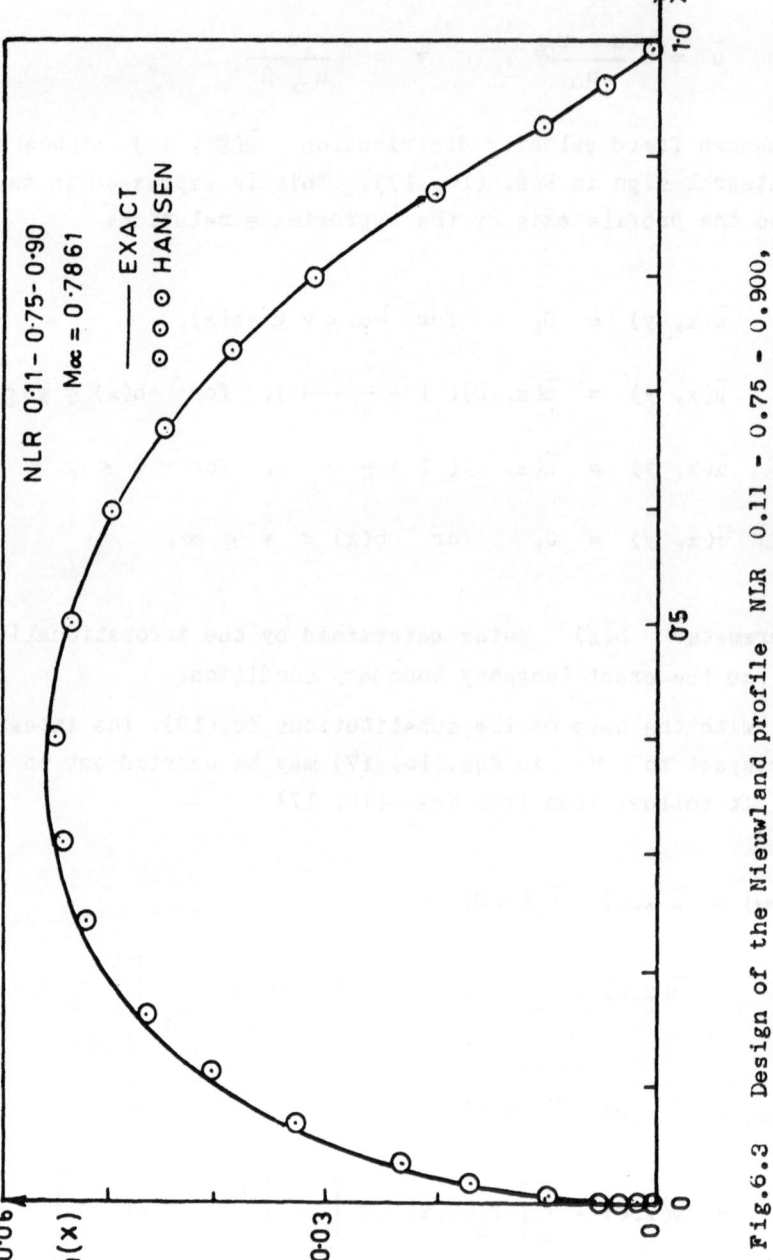

Fig.6.3   Design of the Nieuwland profile NLR 0.11 - 0.75 - 0.900,
$M_\infty = 0.7861$   after Hansen (1976a).  Comparison with the exact solution.

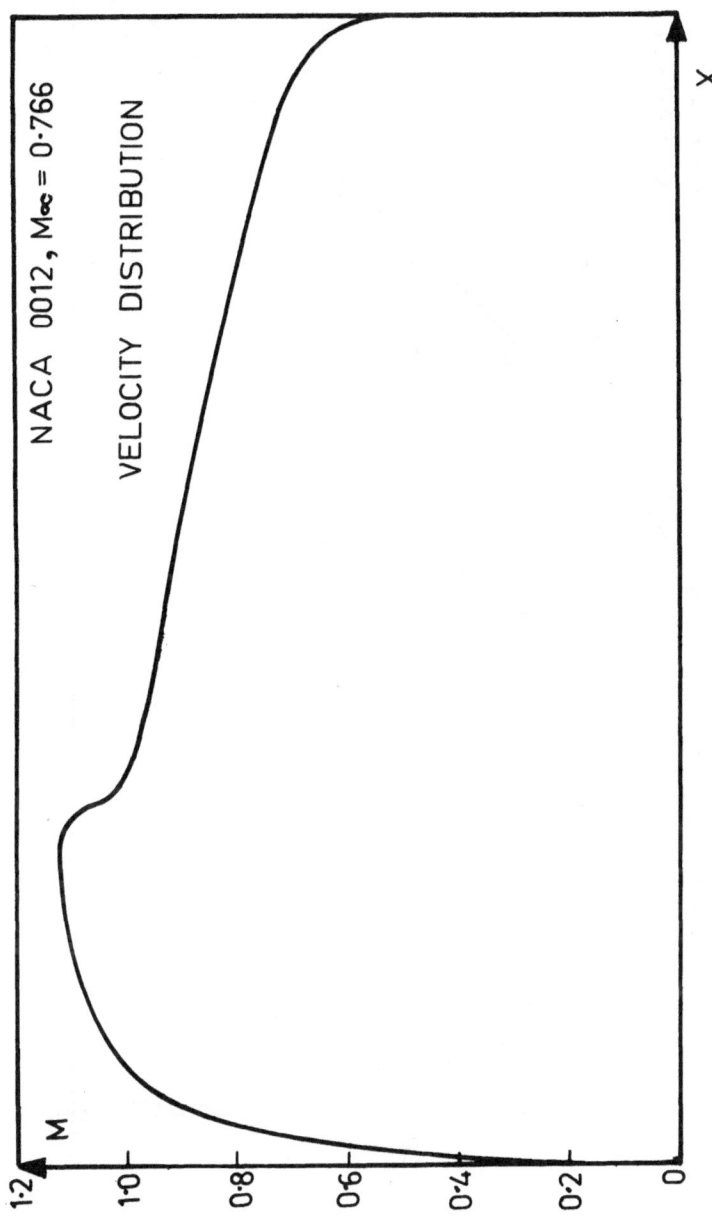

Fig.6.4   Design of a NACA0012 profile at zero incidence with a weak shock at
$M_{\infty} = 0.77$ after Hansen (1975).   Comparison with the solution of Bauer et al.
(1972).   (a)   velocity distribution

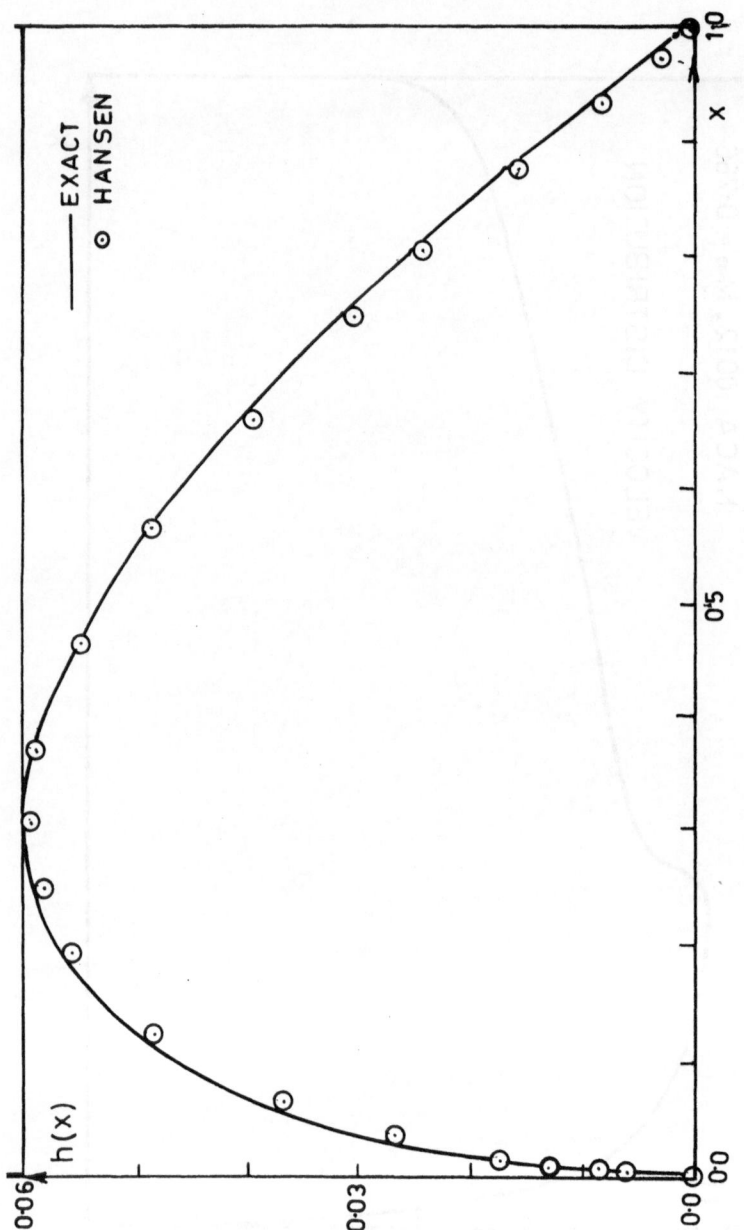

Fig.6.4 Design of a NACA 0012 profile at zero incidence with
a weak shock at $M_\infty = 0.77$ after Hansen (1975).
Comparison with the solution of Bauer et al. (1972).
(b) Profile shape.

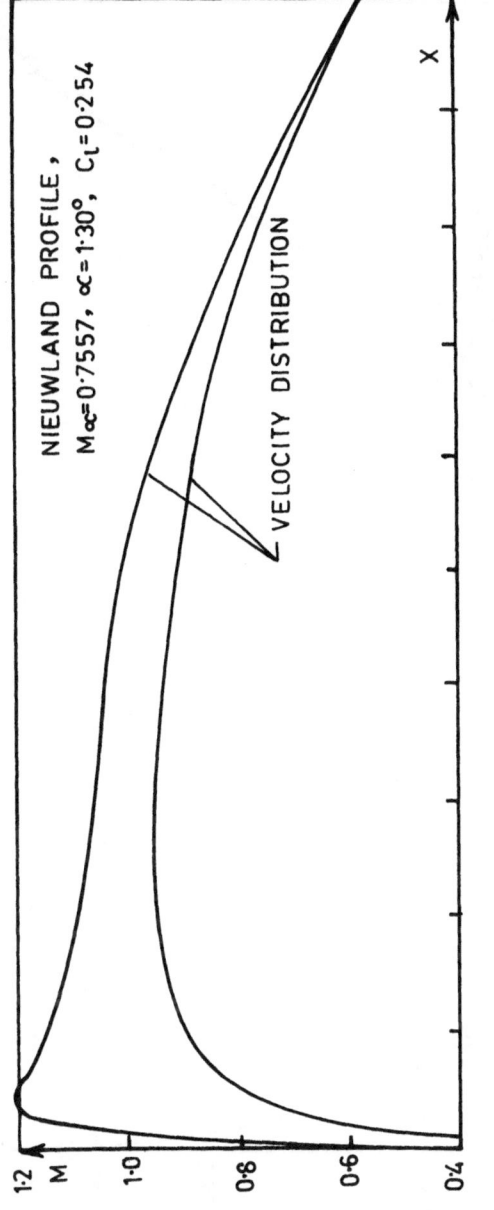

Fig.6.5    Design of a lifting supercritical Nieuwland profile, after Hansen
(1975).   Comparison with exact results taken from Lock (1970).
(a) velocity distribution.

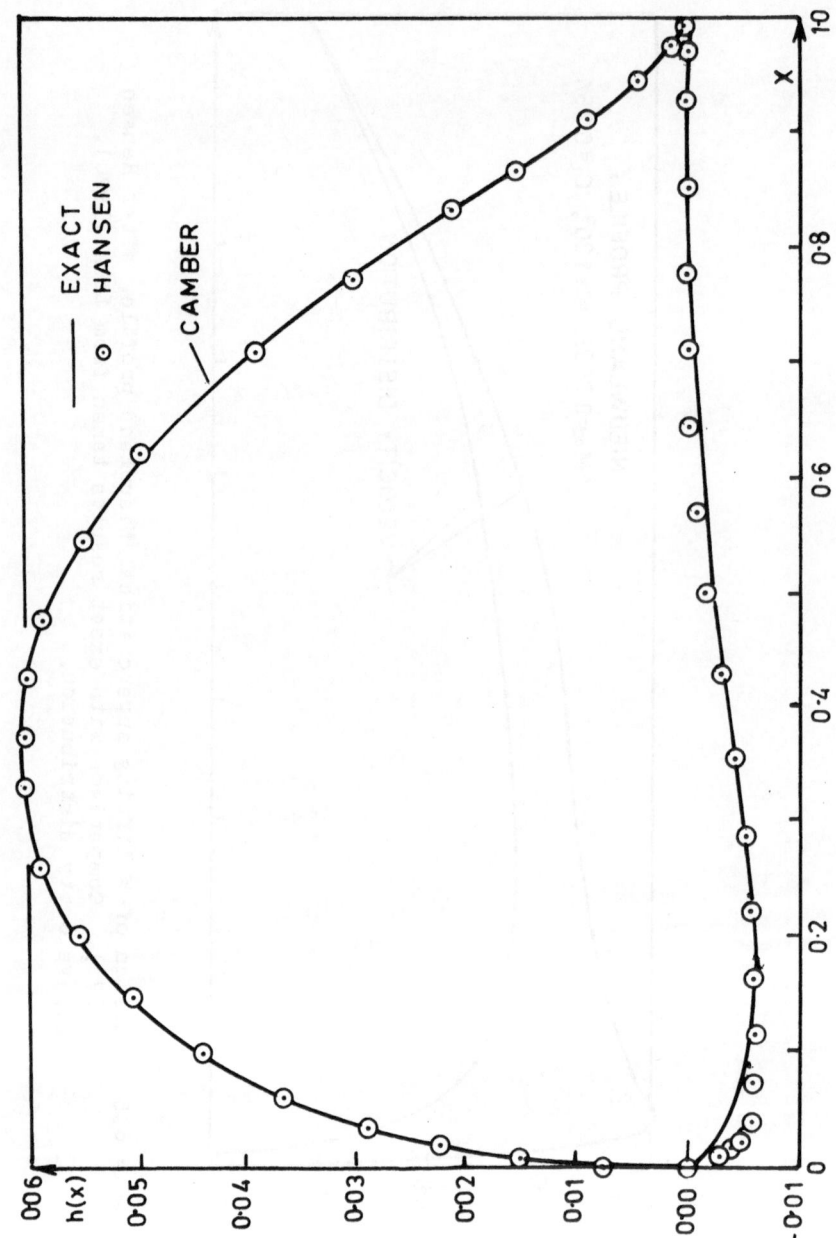

Fig.6.5 Design of a lifting supercritical Nieuwland profile, after Hansen (1975). Comparison with exact results, taken from Lock (1970). (b) Profile shape.

For the design problem, the velocity distribution $v(x)(=u^2+v^2)$ is prescribed on the profile contour $y = y_p$. The free-stream Mach number $M_\infty$ is also prescribed. However, the profile contour and the perturbation velocity components $\bar{u}(x, 0)$, $\bar{v}(x, 0)$ and the parameter $b(x)$ are unknown. Consequently an iteration has to be performed, starting from an initial guess at the profile contour, so that $\bar{u}(x, 0)$, $\bar{v}(x, 0)$ and $b(x)$ may be determined initially. Subsequently, performing the integrations in Eqs. (20, 21), the homogeneous part of the velocity distribution $\bar{u}_H(x, 0)$ and $\bar{v}_H(x, 0)$ are determined. By application of Göthert rule, the linear compressible flow solution may be associated with the corresponding incompressible flow. Standard methods may then be used for the incompressible airfoil design, like the method of Weber (1957) and Truckenbrodt (1951). The iteration procedure converges quite rapidly, and in general 4 and 5 iteration steps are needed for satisfactory convergence.

Fig.6.3 shows the design of a shock-free symmetrical Nieuwland profile at zero incidence computed by Hansen (1976a) with 64 pivotal points on the profile axis. The design procedure converges in only 5 iteration steps. It shows excellent agreement with the exact solution, with less than 2 per cent error. However, the maximum thickness $\tau = 0.114$ could not be exactly reached by the iteration.

Fig.6.4 shows the computational results of Hansen for a NACA 0012 profile at zero incidence with a weak shock, at free-stream Mach number $M_\infty = 0.766$. The contour velocity distribution was calculated using the program of Bauer et al. (1972). In this case also, 64 pivotal points were used and converged solution was obtained in 5 iteration steps. The agreement is excellent. A lifting supercritical shock-free case is presented in Fig.6.5. The thickness and camber distributions are compared with the exact solution taken from Lock (1970). Computations were carried out with 64 pivotal points each on the upper and lower side and convergence is obtained in 4 iteration steps. The results show excellent agreement.

# Chapter VII

## Existence And Uniqueness Question

### 7.1 Introduction

A great upsurge of research activity on transonic flow has taken place in the past decade and much progress has been achieved in recent times in this area. Transonic flow problems are of interest from theoretical standpoint as well as from the practical aerodynamic standpoint of speeding up the performance of subsonic aircrafts. Of much theoretical interest is the question: How many supercritical solutions exist for the direct transonic problem where the free-stream Mach number and the thin profile or wing shape are prescribed? This question has not been answered as yet, even under the premises of an inviscid irrotational steady small perturbation theory. For purely subsonic flow, the answer is given by the well-known work of Bers (1954). The celebrated Non-existence Theorems of Morawetz (1956) dealt with the indirect problem.

The integral equation method is ideally suited for studying the direct transonic problem. Existence and uniqueness of transonic flow past profiles and wings follow directly from the existence and uniqueness of solutions of the relevant integral equations. However, not much work has been done so far in this direction. The work of Schubert and Schleiff (1969) shows the existence and uniqueness of solution for subcritical symmetrical profile flow at zero incidence, which is discussed in the present chapter. Also the relevant works of the present author have been discussed briefly, which deal with the convergence of the direct iteration scheme and simultaneously deliver a proof of existence and uniqueness of solutions.

### 7.2 Convergence of the direct iteration scheme for a symmetric profile at zero incidence

For the direct problem of steady inviscid transonic flow past a thin symmetric profile at zero incidence, the basic equation is the integral equation of Oswatitsch. The alternative form, Eq.(1.79)

$$U(X, Y) = U_p(X, Y) + \frac{1}{4} U^2(X, Y)$$

$$- \frac{1}{2\pi} \int_{-\infty}^{\infty} \int_{-\infty}^{\infty} \frac{1}{2} U^2(\mathcal{E}, \eta) \frac{(\mathcal{E} - X)^2 - (\eta - Y)^2}{\left[(\mathcal{E} - X)^2 + (\eta - Y)^2\right]^2} d\mathcal{E} \, d\eta ,$$

$$\tag{1}$$

is more convenient for studying the question of existence and uniqueness. Here the linearized solution

$$U_p(X, Y) = \frac{1}{\pi} \int_0^1 V_o(\mathcal{E}) \frac{X - \mathcal{E}}{(X - \mathcal{E})^2 + Y^2} d\mathcal{E} , \tag{2}$$

is a known quantity, $V_o(\mathcal{E}) = V(\mathcal{E}, 0)$ being a known quantity determined by the tangency boundary condition at the profile, Eq.(2.3). As observed in section 1.7.2, Eq.(1) is valid for shock-free flow as well as for flows with weak shocks, not necessarily straight and normal. If a shock be present, it should have a small curvature.

For solving Eq.(1), the following direct iteration scheme was introduced in section 3.7:

$$U_{n+1}(X, Y) = U_p(X, Y) + \frac{1}{4} U_n^2(X, Y)$$

$$- \frac{1}{2\pi} \int_{-\infty}^{\infty} \int_{-\infty}^{\infty} U_n^2(\mathcal{E}, \eta) \frac{(\mathcal{E} - X)^2 - (\eta - Y)^2}{\left[(\mathcal{E} - X)^2 + (\eta - Y)^2\right]^2} d\mathcal{E} \, d\eta ,$$

$$n = 0, 1, 2, 3, \ldots \tag{3}$$

In the present section we study (Niyogi 1978b) analytically the convergence of this scheme with the starting solution

$$U_o(X, Y) = U_p(X, Y) \tag{4}$$

by means of the Contraction Mapping Principle (Rall 1969), in the space $L_2(E_2)$. For this purpose Eq.(1) is rewritten in polar coordinates.

Referred to $\bar{X} = (X, Y)$ as pole, let $(r, \theta)$ denote polar coordinates of the point $\bar{Y} = (\xi, \eta)$ of the Eucledian space $E_2$. Then with $U = U(\bar{X})$, Eq.(1) may be put to the form

$$U(\bar{X}) = U_p(\bar{X}) + \frac{1}{4} U^2(\bar{X}) - \frac{1}{2\pi} \int_{E_2} \frac{1}{2} U^2(\bar{Y}) \frac{\cos 2\theta}{r^2} d\bar{Y},$$

(5)

and the iteration scheme, Eqs.(3,4) takes the form

$$U_{n+1}(\bar{X}) = U_p(\bar{X}) + \frac{1}{4} U_n^2(\bar{X}) - \frac{1}{2\pi} \int_{E_2} \frac{1}{2} U_n^2(\bar{Y}) \frac{\cos 2\theta}{r^2} d\bar{Y},$$

$$n = 0, 1, 2, 3, \ldots, \qquad (6)$$

with a suitable starting solution $U_0 = U_p$.

Assuming $U_p(\bar{X})$ and $U(\bar{X})$ square integrable in Lebesgue sense in the Eucledian space $E_2$, we see after Mikhlin (1965), that the operator defined by

$$A(U) = U_p(\bar{X}) + \frac{1}{4} U^2(X) - \frac{1}{2\pi} \int_{E_2} \frac{1}{2} U^2(\bar{Y}) \frac{\cos 2\theta}{r^2} d\bar{Y}, \qquad (7)$$

also belongs to the space $L_2(E_2)$. Mikhlin (1965) considered operators in $L_2(E_2)$ of the form

$$F(Q) = a Q(\bar{X}) + \frac{1}{2\pi} \int_{E_2} \frac{f(\theta)}{r^2} Q(\bar{Y}) d\bar{Y}, \qquad (8)$$

where $a$ is a constant, and established that the singular operator $F(Q)$ is bounded in $L_2(E_2)$ and that

$$\|F\| = \text{ess max} \left| \Phi_F(\theta) \right|, \qquad (9)$$

where $\Phi_F(\theta)$ denotes the symbol of the singular operator $F$ (discussed in Appendix A1).

A little calculation (Niyogi 1973) shows that the symbol of the operator

$$F(Q) \;=\; \tfrac{1}{4}\, Q(\bar{X}) \;-\; \frac{1}{2\pi} \int_{E_2} \frac{Q(\bar{Y})}{2} \; \frac{\cos 2\theta}{r^2}\, d\,\bar{Y} \;, \tag{10}$$

is

$$\Phi_F(\theta) \;=\; \tfrac{1}{4} + \tfrac{1}{4} \cos 2\theta. \tag{11}$$

We see from Eq.(9) and (11) that

$$\|\,F\,\| \;=\; \tfrac{1}{2}\,. \tag{12}$$

We now consider a closed ball $\bar{W}(U_0\,,\,R)$ in $L_2(E_2)$. Noting that for two points $U'\!,\ U'' \in \bar{W}(U_0,\,R)$

$$\|\,U' + U''\,\| \;\le\; 2\,(\bar{U}_0 + R), \tag{13}$$

where $\bar{U}_0$ denotes the $L_2$-norm of $U_0$, we have

$$\|\,A(U') - A(U'')\,\| \;=\; \left\|\, \frac{U'^2 - U''^2}{4} - \frac{1}{2\pi} \int_{E_2} \frac{U'^2 - U''^2}{2} \, \frac{\cos 2\theta}{r^2}\, d\,\bar{Y} \,\right\|$$

$$\le \quad (\bar{U}_0 + R)\,\|\,U' - U''\,\|. \tag{14}$$

Thus the operator $A$ is a contraction mapping of the closed ball $\bar{W}(U_0,\,R)$ in $L_2(E_2)$ with contraction factor $\bar{\theta}$, provided

$$\bar{\theta} \;=\; \bar{U}_0 + R < 1. \tag{15}$$

Further, choosing the starting solution $U_o = U_P$, we have from Eq.(6)

$$U_1 = U_P + \frac{1}{4} U_P^2 - \frac{1}{2\pi} \int_{E_2} \frac{1}{2} U_P^2(\bar{Y}) \frac{\cos 2\theta}{r^2} \, d\bar{Y}.$$

Consequently,

$$\left\| U_1 - U_o \right\| = \left\| \frac{1}{4} U_P^2 - \frac{1}{2\pi} \int_{E_2} \frac{1}{2} U_P^2(\bar{Y}) \frac{\cos 2\theta}{r^2} \, d\bar{Y} \right\|$$

$$\leq \frac{1}{2} (\bar{U}_P)^2. \tag{16}$$

Thus, provided condition (15) is satisfied, we have (Rall 1969)

$$R \geqslant \frac{1}{1-\bar{\theta}} \left\| U_1 - U_o \right\|,$$

from which follows by conditions (15) and (16)

$$R^2 - (1 - \bar{U}_P)R + \frac{1}{2}(\bar{U}_P)^2 \leq 0. \tag{17a}$$

Inequality(17a) has the formal solution

$$\frac{1}{2} \left\{ 1 - \bar{U}_P - \left[ 1 - 2\bar{U}_P - (\bar{U}_P)^2 \right]^{1/2} \right\} < R < \frac{1}{2} \left\{ 1 - \bar{U}_P \right.$$

$$\left. + \left[ 1 - 2\bar{U}_P - (\bar{U}_P)^2 \right]^{1/2} \right\}, \tag{17b}$$

assuming the radicand to be real. The maximum permissible value of $\bar{U}_P$ for which inequality (17b) will be true is determined from the equation

$$1 - 2\bar{U}_P - (\bar{U}_P)^2 = 0,$$

155

to be

$$\left. \bar{U}_P \right)_{max} = \sqrt{2} - 1. \tag{18a}$$

For this value of $\bar{U}_P$ , inequality (17b) gives

$$\left. R \right)_{max} = \frac{1}{2} (2 - \sqrt{2}). \tag{18b}$$

The corresponding value of the contraction factor $\bar{\theta}$ is

$$\bar{\theta} = \sqrt{2}/2 < 1,$$

so that the condition (15) for contraction is satisfied. Thus according to the Contraction Mapping Principle (Rall 1969), unique solution of the basic equation (1) exists in the closed ball $\bar{W}(U_P , R)$, provided the $L_2$-norm of the linearized solution $\bar{U}_P$ satisfies condition (18a). Noting that the norm $\bar{U}_P$ may be expressed in terms of the reduced thickness ratio T of the profile, which is a transonic similarity parameter, Eqs.(18) determine the range of the permissible transonic similarity parameter, for which unique solution exists. Simultaneously, we obtain a proof of convergence of the direct iteration scheme as well as an estimate of the rate of convergence of the scheme. That the condition (18a) for the maximum permissible value of the norm $\bar{U}_P$ actually determines supercritical transonic solution is demonstrated below by means of numerical computations.

The error estimate for the n-th iteration step is given by (Rall 1969)

$$\left\| U_n - U \right\| \le \frac{\bar{\theta}^n}{1 - \bar{\theta}} \left\| U_1 - U_o \right\|.$$

Using Eqs. (15) and (16) we get

$$\left\| U_n - U \right\| \le \frac{(\bar{U}_P + R)^n}{1 - (\bar{U}_P + R)} \cdot \frac{1}{2} (\bar{U}_P)^2. \tag{19a}$$

Corresponding to the maximum permissible value of $\bar{U}_p$ given by Eq.(18a), the error estimate for the n-th iteration step is found on simplification to be

$$\|\, U_n - U \,\| \leq 2^{-(n+1)/2}. \tag{19b}$$

However, for lower values of $\bar{U}_p$ , the rate of convergence would be faster.

## Calculation of the norm

The $L_2$ -norm of the linearized solution is given by

$$\bar{U}_p = \left[ \int_{-\infty}^{\infty} \int_{-\infty}^{\infty} \left\{ U_p(X, Y) \right\}^2 dX\, dY \right]^{1/2}. \tag{20}$$

For the case of a symmetric parabolic arc profile, Eq.(2.15) the linearized solution in reduced coordinates is found as

$$U_p(X, Y) = \frac{T}{\pi} \left[ (2X - 1)\, \ln\left|(X - 1)^2 + Y^2\right| + (1 - 2X)\ln\left|X^2 + Y^2\right| \right.$$

$$\left. + 4 - 4Y \left\{ \tan^{-1} \frac{1 - X}{Y} + \tan^{-1} \frac{X}{Y} \right\} \right].$$

Since the integrations in Eq.(20) cannot be performed in closed form, we evaluate the norm numerically. Noting that $U_p(X, Y)$ decays rapidly like a dipole away from the profile, the integrations need not stretch over the entire infinite plane. We approximately replace it by a finite rectangular domain D around the profile such that the contributions to $\bar{U}_p$ from the region outside D becomes negligibly small. The domain D is subdivided into a rectangular network and the integration is performed by Simpson's $\frac{1}{3}$ -rule of double integration, taking appropriate care for the singularity at (0, 0) and (0, 1), as explained in section 3.7. The computed values of the norm for a parabolic arc profile computed with 100 x 50 mesh points, stretching between $-3 \leq x \leq 3$ , $0 \leq Y \leq 3$, have been presented in the following table:

Table 1:  Calculation of the norm $\bar{U}_p$ for a parabolic arc profile

| Reduced thickness ratio | Norm | Reduced thickness ratio | Norm |
|:---:|:---:|:---:|:---:|
| $T$ | $\bar{U}_p$ | $T$ | $\bar{U}_p$ |
| 0·5 | 0.24 | 0.70 | 0.34 |
| 0.6 | 0.29 | 0.75 | 0.37 |
| 0.65 | 0.32 | 0.80 | 0.39 |
| 0.67 | 0.33 | 0.82 | 0.40 |

Computations with the direct iteration scheme presented in section 3.7 show that the flow is critical for $T = 0.67$ (slightly less than it) as shown in Fig.3.13. From this and from table 1, we see that the norm $\bar{U}_p$ satisfies the condition (18a) for values of the reduced thickness ratio in the range $0.67 \leq T \leq 0.82$. For all these supercritical flows, the direct iteration scheme Eq.(3) converges to the unique exact solution of the integral equation of Oswatitsch. It should be noted that the solutions being Lebesgue square integrable functions may possess jump discontinuities.

The full implications and evaluation of the above results, and in particular, their relationship with the Non-existence Theorems of Morawetz (1956) are at present being studied. The above results may be readily extended to the three-dimensional case which is taken up in the next section (Niyogi 1981).

## 7.3 Flow past a thin symmetric wing at zero incidence

The procedure discussed in the previous section for studying existence and uniqueness of flow past a thin symmetric profile at zero incidence may be readily extended to the three-dimensional case of flow past a thin symmetrical wing at zero incidence. This is undertaken in the present section. The corresponding integral equation, as discussed in section 5.1 is (Eq.(5.8)):

$$U(X,Y,Z) = U_p(X, Y, Z) + \frac{1}{6} U^2(X, Y, Z)$$
$$- \frac{1}{8\pi} \int_{-\infty}^{\infty} \int_{-\infty}^{\infty} \int_{-\infty}^{\infty} \frac{2(\varepsilon-X)^2 - (\eta-Y)^2 - (\zeta-Z)^2}{\left[ (\varepsilon-X)^2 + (\eta-Y)^2 + (\zeta-Z)^2 \right]^{5/2}} U^2(\varepsilon,\eta,\zeta) d\varepsilon d\eta \, d\zeta, \qquad (21)$$

and the corresponding direct iteration scheme is

$$U_{n+1}(X, Y, Z) = U_p(X, Y, Z) + \frac{1}{6} U_n^2(X, Y, Z)$$

$$- \frac{1}{8\pi} \int_{-\infty}^{\infty} \int_{-\infty}^{\infty} \int_{-\infty}^{\infty} \frac{2(\xi-X)^2 - (\eta-Y)^2 - (\zeta-Z)^2}{\left[ (\xi-X)^2 + (\eta-Y)^2 + (\zeta-Z)^2 \right]^{5/2}} U_n^2(\xi,\eta,\zeta) \, d\xi \, d\eta \, d\zeta,$$

$$n = 0, 1, 2, 3, \ldots , \quad (22)$$

with a suitable starting solution

$$U_o = U_o(X, Y, Z).$$

We study here the convergence of the direct iteration scheme (22) analytically, by means of Banach Contraction Mapping Principle (Rall 1969) in the space $L_2(E_3)$.

For our purpose it appears convenient to rewrite Eq.(22) in spherical polar coordinates. With $\bar{X} = (X, Y, Z)$ as pole, let $(r, \theta, \emptyset )$ denote the spherical polar coordinates of the point $\bar{Y}(\xi, \eta , \zeta)$ of the Eucledian space $E_3$. Then with $U = U(\bar{X})$, Eq.(21) may be put to the form

$$U(\bar{X}) = U_p(\bar{X}) + \frac{1}{6} U^2(\bar{X}) - \frac{1}{4\pi} \int_{E_3} U^2(\bar{Y}) \frac{f(\theta)}{r^3} \, d\bar{Y} , \quad (23a)$$

where

$$f(\theta) = \frac{1}{2} (3 \cos^2 \theta - 1), \quad (23b)$$

and the iteration scheme, Eq.(22) takes the form

$$U_{n+1}(\bar{X}) = U_p(\bar{X}) + \frac{1}{6} U_n^2 (\bar{X}) - \frac{1}{4\pi} \int_{E_3} U_n^2 (\bar{Y}) \frac{f(\theta)}{r^3} \, d\bar{Y} ,$$

$$n = 0, 1, 2, 3, \ldots \quad (24)$$

Mikhlin (1965) considered operators in $L_2(E_3)$ of the form

$$F(Q) = a\ Q(\overline{X}) - \frac{1}{(2\pi)^{3/2}} \int_{E_3} \frac{\overline{f}(\theta)}{r^3}\ Q(\overline{Y})\ d\overline{Y} , \qquad (25)$$

where $a$ is a constant and the characteristic $\overline{f}(\theta)$ satisfies the condition

$$\int_S \overline{f}(\theta)\ dS = 0, \qquad (26)$$

where $S$ is the unit sphere over which $\theta$ moves, which holds in the present case for $f(\theta)$ defined by Eq.(23b). Mikhlin established that the operator $F(Q)$ is bounded in $L_2(E_3)$ and as in the two-dimensional case

$$||F|| = \text{ess max } \left|\Phi_F(\theta)\right| \qquad (27)$$

where $\Phi_F(\theta)$ is the symbol of the singular operator $F$.

Eq.(23a) may be rewritten as an operator equation

$$U = A(U) \qquad (28a)$$

where the operator

$$A(U) = U_p(\overline{X}) + \sqrt{\frac{\pi}{2}}\ \left[\ \frac{U^2(\overline{X})}{3\sqrt{2\pi}} - \frac{1}{(2\pi)^{3/2}} \int_{E_3} \frac{f(\theta)}{r^3}\ U^2(\overline{Y})\ d\overline{Y}\ \right].$$

$$(28b)$$

The symbol of the singular operator $F$ in Eq.(25), which is defined in terms of the Fourier transform of the singular kernel, has been calculated by A.K. Niyogi (1976) for the characteristic function $f(\theta)$ defined by Eq.(23b), as (Appendix A2)

$$\Phi_F = a - \frac{1}{\sqrt{2\pi}}\ \sin^2\theta. \qquad (29)$$

Consequently, in view of Eq.(27), the $L_2$-norm of the operator $A$ in Eq.(28b) is

$$\left|\left| A \right|\right| = \sqrt{\frac{\pi}{2}} \cdot \frac{1}{3 \sqrt{2\pi}} = \frac{1}{6} \cdot \qquad (30)$$

We consider now two points $U'$ and $U''$ in the closed ball $\bar{W}(U_o, R)$ in $L_2(E_3)$. Then noting that

$$\left|\left| U' + U'' \right|\right| \leq 2(\bar{U}_o + R), \qquad (31)$$

where $\bar{U}_o$ is the $L_2$-norm of $U_o$, we have

$$\left|\left| A(U') - A(U'') \right|\right|$$

$$= \sqrt{\frac{\pi}{2}} \left|\left| \frac{U'^2 - U''^2}{3 \sqrt{2\pi}} - \frac{1}{(2\pi)^{3/2}} \int_{E_3} \frac{f(\theta)}{r^3} (U'^2 - U''^2) \, d\bar{Y} \right|\right|$$

$$\leq \frac{1}{6} \cdot 2(\bar{U}_o + R) \left|\left| U' - U'' \right|\right|. \qquad (32)$$

Consequently, the operator $A$ in $L_2(E_3)$ defined by Eq.(28b) is a contraction mapping of the closed ball $\bar{W}(U_o, R)$, provided the contraction factor $\bar{\theta}$ satisfies the condition

$$\bar{\theta} = \frac{1}{3}(\bar{U}_o + R) < 1. \qquad (33)$$

Further, choosing the starting solution as $U_o = U_p$, we have from Eq.(22)

$$U_1 = U_p + \frac{1}{6} U_p^2 - \frac{1}{4\pi} \int_{E_3} U_p^2 \frac{f(\theta)}{r^3} \, d\bar{Y}.$$

Therefore,

$$\left|\left| U_1 - U_o \right|\right| = \left|\left| \frac{1}{6} U_p^2 - \frac{1}{4\pi} \int_{E_3} U_p^2 \frac{f(\theta)}{r^3} \, d\bar{Y} \right|\right|$$

$$\leq \frac{1}{6}(\bar{U}_p)^2. \qquad (34)$$

Thus, provided condition (33) is satisfied, we have (Rall 1969)

$$R \gg \frac{1}{1 - \bar{\theta}} \left\| U_1 - U_o \right\| ,$$

from which follows by conditions (33) and (34)

$$R \gg \frac{1}{1 - \frac{1}{3}(\bar{U}_P + R)} \frac{(\bar{U}_P)^2}{6} . \tag{35}$$

On simplification, this yields

$$R^2 - (3 - \bar{U}_P) R + \frac{1}{2}(\bar{U}_P)^2 \leq 0. \tag{36}$$

Inequality (36) has the formal solution

$$\frac{1}{2}\left\{ 3 - \bar{U}_P - \left[(3 - \bar{U}_P)^2 - 2(\bar{U}_P)^2\right]^{1/2}\right\} < R < \frac{1}{2}\left\{ 3 - \bar{U}_P \right.$$

$$\left. + \left[(3 - \bar{U}_P)^2 - 2(\bar{U}_P)^2\right]^{1/2}\right\} . \tag{37}$$

The maximum value of $\bar{U}_P$ for which inequality (37) will be true is determined from

$$(3 - \bar{U}_P)^2 - 2(\bar{U}_P)^2 = 0,$$

to be

$$\left. \bar{U}_P \right)_{max} = 3(\sqrt{2} - 1). \tag{38}$$

For this value of $\bar{U}_P$, inequality (37) delivers

$$\left. R \right)_{max} = 3( 1 - \frac{\sqrt{2}}{2} ).\tag{39}$$

The corresponding value of the contraction factor $\bar{\Theta}$ is

$$\bar{\Theta} = \frac{1}{3}\left[ 3( \sqrt{2} - 1) + 3(1 - \frac{\sqrt{2}}{2} ) \right] = \frac{\sqrt{2}}{2} < 1,\tag{40}$$

so that the condition (33) is satisfied.

Thus according to the Banach Contraction Mapping Principle (Rall 1969), the direct iteration scheme, Eq.(22) converges to the exact solution of Eq.(21) in the closed ball $\bar{W}(U_p , R)$ and the solution is unique there, provided the $L_2$-norm of the linearized solution $\bar{U}_p$ satisfies the condition (38). Eqs.(38) and (39) determine the maximum permissible value of the transonic similarity parameter for which unique solution exists.

In the next section, we present the work of Schubert and Schleiff (1969) on zero incidence profile flow.

## 7.4 Schubert-Schleiff series

For shock-free flow past a thin symmetric profile at zero incidence, Schubert and Schleiff (1969) proved the existence and uniqueness of solution for subcritical flow, which is discussed in the present section.

We start from the alternative form of the integral equation of Oswatitsch, Eq.(1.79):

$$U(X, Y) = U_p(X, Y) + \frac{1}{4} U^2(X, Y)$$

$$- \frac{1}{4\pi} \int_{-\infty}^{\infty} \int_{-\infty}^{\infty} U^2(\mathcal{E}, \eta) \frac{(\mathcal{E} - X)^2 - (\eta - Y)^2}{\left[(\mathcal{E} - X)^2 + (\eta - Y)^2\right]^2} d\mathcal{E}\, d\eta ,\tag{41}$$

where the linearized solution is

$$U_p(X, Y) = \frac{1}{\pi} \int_0^1 V_0(\mathcal{E}) \frac{X - \mathcal{E}}{(X - \mathcal{E})^2 + Y^2} d\mathcal{E}. \tag{42}$$

Here $V_0(\mathcal{E}) = V(\mathcal{E}, 0)$ is a known quantity, determined by the tangency boundary condition at the profile. The two dimensional Eucledian space is denoted by $E_2$. It is assumed that $U_p(X, Y)$ is a Hölder continuous function and consequently, as proved by Schubert and Schleiff (1969), it belongs to the space $L_4( E_2 )$. We seek solution $U(X, Y)$ of Eq.(41) in the space of bounded and uniformly Hölder continuous functions in $E_2$ which also belong to $L_4(E_2)$. After Vekua (1963) the norm is defined by

$$L_4 C_\alpha (U, E_2) = ( \int\int_{E_2} |U|^4 \, dX \, dY )^{1/4} + \operatorname*{Sup}_{E_2} |U|$$

$$+ \operatorname*{Sup}_{E_2} \left| \frac{U(Z + h) - U(Z)}{h^\alpha} \right|. \tag{43}$$

We now write Eq.(1) in the form

$$U = A \{ U^2 \} + U_p. \tag{44}$$

According to Vekua (1963, p.54), the operator $A$ is a bounded operator in $L_4 C_\alpha (E_2)$. Therefore

$$L_4 C_\alpha (A \{ U^2 \} ) \leq M L_4 C_\alpha(U^2). \tag{45}$$

The operator $U^2$ is a bounded operator and certainly its norm can be made arbitrarily small, because

$$L_4 C_\alpha ( U^2 ) \leq 2 L_4 C_\alpha (U) L_4 C_\alpha(U), \tag{46}$$

if the norm of U in $L_4 C_\alpha$ be sufficiently small. If we choose the norm of U sufficiently small, the operator $A\{U^2\}$ will contract in $L_4 C_\alpha$. For small U and small $U_p$ , which means small perturbation $V_o(X)$ , the right hand side of Eq.(44) maps into a sphere in the space $L_4 C_\alpha$ .

According to the principle of Contraction Mapping, Eq.(44) possesses exactly one solution, which may be found by iteration. For example, it may be represented as the following series

$$U = U_p + A\{U_p^2\} + A\{2 U_p A\{U_p^2\}\} + \ldots , \qquad (47)$$

For studying the continuation possibility of the solution for growing $U_p$ , we consider the derivative of the implicit operator

$$P(U_p, U) = U - A\{U^2\} - U_p. \qquad (48)$$

The Fréchet derivative at the point $U_o$ is

$$P'_{U_o}\{U\} = U - 2A\{U_o U\}. \qquad (49)$$

For $U_p = 0$, the equation

$$P(U_p, U) = 0, \qquad (50)$$

has the solution $U \equiv 0$ . At this position $U_o = 0$, $P'_{U_o} = U$, is the identity operator which possesses a bounded inverse operator. Therefore Eq.(50) is uniquely solvable in U and Eq.(47) represents the only continuous solution starting from $U_o$ . Further unique continuation of the solution is possible as long as $P'_{U_o}\{U\}$ is invertible. We rewrite now Eq.(49) in the form

$$P'_{U_o} \left\{ U \right\} = U(X, Y) - \frac{1}{2} U_o(X, Y) \left[ U(X, Y) - \frac{1}{\pi} \iint_{E_2} U(\xi,\eta) \frac{\cos 2\phi}{r} dr \, d\phi \right]$$

$$+ \frac{1}{2\pi} \iint_{E_2} U(\xi,\eta) \frac{U_o(\xi,\eta) - U_o(X,Y)}{r} \cos 2\phi \, dr \, d\phi, \qquad (51)$$

and investigate this operator with the theory of Mikhlin (1965). The last integral on the right of Eq.(51) represents a completely continuous operator, and consequently delivers no contribution to the symbol (Mikhlin 1965) of the singular integral operator $P'_{U_o} \left\{ U \right\}$. The symbol is thus (see appendix A1):

$$\sigma(X, Y; \phi) = 1 - U_o(X, Y) \frac{1 + \cos 2\phi}{2}. \qquad (52)$$

If there be points where $U_o(X, Y) = 1$, then the symbol vanishes at individual points and $P'_{U_o}$ can no more possess bounded inverse operator. Therefore solution of Eq.(41) in general cannot be continued uniquely as soon as with increasing $U_p$ values, points with $U = 1$ occur. Hence the Schubert Schleiff series Eq.(47) converges to the unique exact solution of Eq.(41) for subcritical flow.

## 7.5  Shock-free flow past thin wings

Existence and uniqueness of shock-free flow past thin wings may also be studied by the procedure of the previous section. Such a study has been made by the present author in the work Niyogi (1981, unpublished) and the present discussion is based on this work.

We assume the known function $U_p(X, Y, Z)$ uniformly Lipschitz continuous with index $\alpha$ in any finite part of $E_3$ and at infinity

$$U_p(X, Y, Z) = O(R^{-\lambda}), \quad \lambda > 0, \quad R^2 = X^2 + Y^2 + Z^2.$$

$$(53)$$

Then it may be readily seen that it also belongs to the space $L_4(E_3)$. We seek solution $U(X, Y, Z)$ of Eq.(21) in the space of bounded and uniformly Lipschitz continuous functions in $E_3$ which also belong to $L_4(E_3)$. These functions constitute a Banach space. Following Schubert and Schleiff (1969), the norm is defined by

$$L_4 \, C_\alpha(U, E_3) = \left( \iiint_{E_3} |U|^4 \, dX \, dY \, dZ \right)^{1/4} + \sup_{E_3} |U| +$$

$$+ \quad \text{Sup}_{E_3} \left| \frac{U(\bar{X}) - U(\bar{Y})}{r^\alpha} \right| , \tag{54}$$

where $\bar{X} = (X, Y, Z)$, $\bar{Y} (= (\mathcal{E}, \eta, \varsigma)$ and $(r, \theta, \emptyset)$ denote the spherical polar coordinates of $\bar{Y}$ with respect to the pole $\bar{X}$.

Rewriting Eq.(21) in the form

$$U = A\{U^2\} + U_p , \tag{55}$$

it is easy to see that for sufficiently small values of $U$ the operator $A\{U^2\}$ will contract in $L_4 \, C_\alpha$ and unique existence of solution is assured by Banach Contraction Mapping Principle.

For studying the continuation possibility to the solution for growing $U_p$, we consider the derivative of the implicit operator

$$P(U_p, U) = U - A\{U^2\} - U_p . \tag{56}$$

The Fréchet derivative at the point $U_o$ is

$$P'_{U_o}\{U\} = U - 2A\{U_o \, U\} . \tag{57}$$

For $U_p \equiv 0$, the equation

$$P(U_p, U) = 0, \tag{58}$$

has the solution $U \equiv 0$. At this position $U_o = 0$, $P'_{U_o} = U$ is the identity operator which possesses a bounded inverse operator. Therefore, Eq. (55) is uniquely solvable in $U$. Further, unique continuation of the solution is possible so long as $P'_{U_o} \{ U \}$ is invertible.

We rewrite Eq.(57) in the form

$$P'_{U_o} \{ U \} = U - 2U_o \left[ \frac{U}{6} - \frac{1}{4\pi} \int_{E_3} \frac{f(\theta)}{r^3} U \, d\bar{Y} \right]$$

$$+ \frac{1}{2\pi} \int_{E_3} U(\xi,\eta,\zeta) \frac{U_o(\xi,\eta,\zeta) - U_o(X,Y,Z)}{r^3} f(\theta) \, d\bar{Y} ,$$

(59)

where the characteristic $f(\theta)$ is defined in Eq.(23b).

The invertibility of the singular operator $P'_{U_o} \{ U \}$ is investigated with the theory of Mikhlin (1965), according to which it is invertible if its symbol is finite with exact lower bound positive, the symbol being defined in terms of the Fourier transform of the singular kernel and discussed in Appendix A2.

We note that the last integral in Eq.(59) represents a completely continuous operator and consequently delivers no contribution to the symbol of the singular integral operator $P'_{U_o} \{ U \}$. The symbol of the singular operator within the square bracket in Eq.(59) has been calculated by A.K. Niyogi (1976) (also Appendix A2) to be equal to

$$\sqrt{\frac{\pi}{2}} \left[ \frac{1}{3\sqrt{2\pi}} - \frac{1}{\sqrt{2\pi}} \sin^2 \theta \right] .$$

(60)

Using this value, the symbol of the operator $P'_{U_o} \{ U \}$ is found on simplification to be

$$1 - \frac{U_o}{3} + U_o \sin^2 \theta .$$

(61)

Thus in the interval $1 < U_o < 3$, the symbol is finite with exact lower bound positive and consequently according to Mikhlin (1965), the operator $P'_{U_o}$ possesses a bounded inverse operator.

Therefore solution of Eq.(21) can be continued uniquely so long as with increasing $U_P$ values, points with $U_o \geqslant 3$ do not occur. Hence continuous supercritical unique solution of our problem exists so long as the local value of the reduced velocity $U(X,Y,Z)$ at any point of the flow field remains less than the supercritical value 3. This resolves the "Transonic Controversy" (Bers 1958) in favour of existence of shock-free flow.

From the above discussion and Section 7.3 it is to be seen that for a prescribed thin wing shape at zero incidence and prescribed free-stream Mach number, characterized by a supercritical transonic similarity parameter, only one supercritical transonic flow can exist, ensuring uniqueness of the solution. It should be noted that the assumption on the known function is a little unconventional since $U_P(X, Y, Z)$ for conventional wings possesses singularities at the leading and trailing edges. However, if by appropriate design of the wing, it becomes possible to remove these singularities, shock-free flow may be realized.

## A1. Exact solution of a two-dimensional linear singular integral equation with dipole singularity of the kernel

We consider here the exact solution of a linear two-dimensional singular integral equation whose kernel possesses a dipole singularity:

$$Q(x,y) + \frac{1}{2\pi} \int_{-\infty}^{\infty} \int_{-\infty}^{\infty} \frac{(\xi-x)^2 - (\eta-y)^2}{\left[(\xi-x)^2 + (\eta-y)^2\right]^2} Q(\xi,\eta) d\xi \, d\eta = g(x,y),$$

(A1)

where $Q(x,y)$ is the unknown function and $g(x,y)$ is known, whose kernel has a dipole singularity at $\xi = x$, $\eta = y$. The exact solution of this equation may be obtained in two different ways (Niyogi 1973) which we discuss here briefly.

In the first procedure, a result due to Tricomi is used. While studying the product of two-dimensional singular integral operators, Tricomi (1926) observed that the solution of equations like Eq.(A1) could be found if it would be possible to solve an associated one-dimensional singular integral equation.

Tricomi observed that the solution of the integral equation

$$a \, \omega(\bar{x}) + \int_{E_2} \frac{f_1(\theta)}{r^2} \omega(\bar{y}) \, d\bar{y} = h(\bar{x}),$$

(A2)

where $h(\bar{x})$ is a known function, and $\bar{x}$ and $\bar{y}$ are points of the Eucledian space $E_2$, $r$ and $\theta$-polar coordinates of the point $\bar{y}$, with reference to the pole $\bar{x}$, and $a$ = constant, is given by

$$\omega(\bar{x}) = b \, h(\bar{x}) + \int_{E_2} \frac{f_2(\theta)}{r^2} h(\bar{y}) \, d\bar{y}.$$

(A3)

The quantities $b$ and $f_2(\theta)$ of Eq.(A3) are to be determined by the relations (Mikhlin 1965):

$$ab + \alpha = 1, \tag{A4}$$

$$\alpha = 2\pi \int_{-\pi}^{\pi} \bar{f}_1(\Psi) \, \bar{f}_2(\Psi + \pi) \, d\Psi, \tag{A5}$$

and

$$b \, f_1(\theta) + a \, f_2(\theta)$$

$$+ \frac{d}{d\theta} \int_{-\pi}^{\pi} \left\{ \bar{f}_1(\Psi) \, \bar{f}_2(\theta) + \bar{f}_1(\theta) \, \bar{f}_2(\Psi) - \bar{f}_1(\Psi) \bar{f}_2(\Psi + \pi) \right\}.$$

$$\cdot \cot(\Psi - \theta) \, d\Psi = 0, \tag{A6}$$

and $\bar{f}_j(\theta)$, $j = 1,2$ denote indefinite integrals of $f_j(\theta)$.

A necessary and sufficient condition for the existence of the singular double integral in Eq.(A2) is that

$$\int_{-\pi}^{\pi} f_2(\Psi) \, d\Psi = 0, \tag{A7}$$

and the $h(\bar{y})$ should satisfy a Hölder condition with a positive exponent. Eq.(A6) is the associated one-dimensional singular integral equation for the unknown $f_2(\Psi)$.

Noting that in the present case,

$$f_1(\theta) = \cos 2\theta, \quad \bar{f}_1(\theta) = \frac{1}{2} \sin 2\theta, \quad a = 2\pi, \quad h = 2\pi g, \tag{A8}$$

Eqs.(A4,5,8) deliver the following relations for determining $b$ and $f_2(\theta)$ :

$$2\pi b + \pi \int_{-\pi}^{\pi} \bar{f}_2(\Psi + \pi) \sin 2\Psi \, d\Psi = 1, \tag{A9a}$$

and

$$b \cos 2\theta + 2\pi \, f_2(\theta) + \frac{1}{2} \frac{d}{d\theta} \int_{-\pi}^{\pi} \left\{ \bar{f}_2(\theta) \sin 2\Psi + \bar{f}_2(\Psi) \sin 2\theta \right.$$

$$\left. - \bar{f}_2(\Psi + \pi) \sin 2\Psi \right\} \cot(\Psi - \theta) \, d\Psi = 0.$$

(A9b)

The trigonometric functions appearing in Eqs.(A9) suggest to seek solution $f_2(\theta)$ in the class of periodic functions, with period $\pi$. So we assume

$$\bar{f}_2(\Psi + \pi) = \bar{f}_2(\Psi).$$

(A10)

Then observing that

$$\int_{-\pi}^{\pi} \sin 2\Psi \cot(\Psi - \theta) \, d\Psi = 2\pi \cos 2\theta,$$

Eq.(A9b) may be rewritten after elementary trigonometric simplification as

$$b \cos 2\theta + 2\pi \, f_2(\theta) + \pi \cos 2\theta \, f_2(\theta)$$

$$= \sin 2\theta \left\{ 2\pi \bar{f}_2(\theta) - \lambda \right\},$$

(A11)

where $\lambda$ is the constant

$$\lambda \equiv \int_{-\pi}^{\pi} \bar{f}_2(\Psi) \, d\Psi.$$

Differentiating Eq.(A11) with respect to $\theta$ and eliminating $\lambda$ by means of Eq.(A11), a first order linear differential equation for $f_2(\theta)$ is obtained, which when solved and treated simultaneously with Eq.(A8) yields as solution of Eqs.(9)

$$f_2(\theta) = -\frac{1}{\sqrt{3}\ \pi^2}\frac{2\cos 2\theta + 1}{(2 + \cos 2\theta)^2}\ , \qquad b = \frac{1}{\sqrt{3}\ \pi}\ .$$

$$(A12)$$

We observe that Eq.(A12) satisfies condition (A7). Substituting Eqs. (A12,8) in Eq.(A3) and returning back to Cartesian coordinates, it follows as solution of Eq.(A1)

$$Q(x,y) = \frac{2}{\sqrt{3}}\ g(x,y) - \frac{2}{\sqrt{3}\ \pi}\ \int\limits_{-\infty}^{\infty}\int\limits_{-\infty}^{\infty}\ g(\xi,\eta)\frac{3(x-\xi)^2 - (y-\eta)^2}{\left[3(x-\xi)^2 + (y-\eta)^2\right]^2}\ d\xi\ d\eta.$$

$$(A13)$$

It was shown by Schubert (1973) that the above closed form solution Eq.(A13) may be derived from the more general solution of Wolfersdorf (1962). However, the above derivation is shorter.

An alternative approach for deriving the above solution is to use the Fourier transform method, which is quite instructive and particularly convenient for obtaining closed form solution of similar equations. This approach is indicated in the following.

Referred to $\bar{x} = (x,y)$ as pole, let $r$ and $\theta$-denote polar coordinates of the point $\bar{y} = (\xi,\eta)$ of the Eucledian space $E_2$. Then Eq.(A1) may be rewritten as

$$AQ \equiv Q(\bar{x}) + \frac{1}{2\pi}\ \int\limits_{E_2}\ \frac{\cos 2\theta}{r^2}\ Q(\bar{y})\ d\bar{y}\ =\ g(\bar{x}). \qquad (A14)$$

Existence of Fourier transform of singular integrals like that appearing in Eq.(A14) was established by Calderon and Zygmund (1952), so that the operator A may be represented as

$$AQ \equiv F^{-1}\ \Phi\ F\ Q\ , \qquad (A15)$$

where F denotes the Fourier transform in two dimensions, defined by

$$F \, Q \equiv \lim_{\substack{\epsilon \to 0 \\ N \to \infty}} \frac{1}{2\pi} \int_{\epsilon < |\bar{x}| < N} e^{-i(\bar{y} \cdot \bar{z})} Q(\bar{y}) \, d\bar{y} , \qquad (A16)$$

and the symbol $\Phi$ is defined by

$$\Phi \equiv 1 + F\left( \frac{\cos 2\Theta}{r^2} \right). \qquad (A17)$$

This leads to the solution of Eq.(A16). For this purpose, let us first evaluate $\Phi$. It follows from the definition of Fourier transform rewritten in polar coordinates, with $\rho$ and $\gamma$ denoting polar coordinates of $\bar{z}$ with respect to $\bar{x}$ as pole, and from Eq.(A17) that

$$\Phi = 1 + \lim_{\substack{\epsilon \to 0 \\ N \to \infty}} \int_0^{2\pi} \int_\epsilon^N e^{-i\rho r \cos (\Theta - \gamma)} \frac{\cos 2\Theta}{r^2} r \, d\Theta \, dr.$$

Putting $\rho r = t$ in it and noting that

$$\int_0^{2\pi} e^{-it \cos \Theta} \cos 2\Theta \, d\Theta = -2\pi J_2(t), \qquad (A18)$$

where $J_2$ denotes Bessel function of order two, it follows

$$\Phi = 1 + \frac{1}{2\pi} \int_0^\infty \frac{dt}{t} \left[ \cos 2\gamma \left\{ - 2\pi J_2(t) \right\} \right] = 1 - \frac{1}{2} \cos 2\gamma .$$

$$(A19)$$

Observing that $\Phi$ is finite, with exact lower bound positive, it follows (Mikhlin 1965) that if $g(\bar{x}) \in L_2(E_2)$, Eq.(A16) has a solution which is unique in space $L_2(E_2)$ and this solution is expressed by the formula

$$Q = F^{-1} \left[ \Phi \right]^{-1} F g = F^{-1} \left[ \frac{2}{2 - \cos 2\gamma} \right] G , \qquad (A20)$$

where   G   is the Fourier transform of  g .

Faltung theorem may be employed to simplify Eq.(A20).  To do this, the inverse Fourier transform of  $2/(2 - \cos 2\gamma)$  is needed. However, this becomes unbounded.  We consider it to be the limit of the expression  $I_\varepsilon$ ,  as  $\varepsilon \longrightarrow 0$ ,  where

$$I_\varepsilon = \frac{1}{2\pi} \int_0^{2\pi} \int_0^\infty e^{-\varepsilon\rho} \, e^{i\rho r \cos(\gamma-\theta)} \, \rho \, d\rho \, \frac{2}{2 - \cos 2\gamma} \, d\gamma, \; \varepsilon > 0,$$

$$= \frac{1}{\pi} \frac{\partial}{\partial r} \int_0^\pi \frac{d\gamma}{2 - \cos 2\gamma} \int_0^\infty \frac{e^{i\rho r\cos(\gamma-\theta)-\varepsilon\rho}}{i \cos(\gamma - \theta)} \, d\rho .$$

Performing the integration above, follows after some simplification

$$\lim_{\varepsilon \to 0} I_\varepsilon = -\frac{4}{\sqrt{3}} \frac{3\cos^2\theta - \sin^2\theta}{(3\cos^2\theta + \sin^2\theta)^2} + \lim_{\varepsilon \to 0} \frac{2}{3\cos^2\theta + \sin^2\theta} \frac{\varepsilon}{(\varepsilon^2 + r^2)^{3/2}}.$$

(A21)

Because of strong singularity at   $r = 0$ , the limit  $\varepsilon \longrightarrow 0$  cannot be taken immediately in the second term on the right.  Then using Faltung theorem and returning back to Cartesian coordinates, it follows from Eqs. (A20, A21)

$$Q(x,y) = \frac{1}{2\pi} \int_{-\infty}^\infty \int_{-\infty}^\infty \frac{-4}{\sqrt{3}} \frac{3(x-\xi)^2 - (y-\eta)^2}{\left[3(x-\xi)^2 + (y-\eta)^2\right]^2} \, g(\xi,\eta) \, d\xi \, d\eta$$

$$+ \lim_{\varepsilon \to 0} \frac{1}{2\pi} \cdot 2\varepsilon \int_{-\infty}^\infty \int_{-\infty}^\infty \frac{(x-\xi)^2 + (y-\eta)^2}{3(x-\xi)^2 + (y-\eta)^2} \frac{g(\xi,\eta) \, d\xi \, d\eta}{\left[\varepsilon^2 + (x-\xi)^2 + (y-\eta)^2\right]^{3/2}}.$$

(A22)

Noting that the only non-zero contribution to the second integral on the right of Eq.(A22) can come from the singularity at $\mathcal{E} = x$, $\eta = y$, $g(\mathcal{E}, \eta)$ in Eq.(A22) may be replaced by $g(x, y)$ and the integration may be carried out. Simplifying we obtain the unique solution of Eq. (A1) as

$$Q(x,y) = \frac{2}{\sqrt{3}} g(x,y) + \frac{2}{\sqrt{3}\,\pi} \int_{-\infty}^{\infty} \int_{-\infty}^{\infty} \frac{(y-\eta)^2 - 3(x-\mathcal{E})^2}{\left[(y-\eta)^2 + 3(x-\mathcal{E})^2\right]^2} \, d\mathcal{E} \, d\eta ,$$

(A23)

which is identical with the solution (A13).

An integral equation, very similar to Eq.(A1) was solved by Niyogi (1969) using the first approach. We present the result as an inversion formula:

$$Q(x,y) - \frac{1}{2\pi} \int_{-\infty}^{\infty} \int_{-\infty}^{\infty} \frac{(\mathcal{E}-x)^2 - (\eta - y)^2}{\left[(\mathcal{E}-x)^2 + (\eta - y)^2\right]^2} Q(\mathcal{E}, \eta) \, d\mathcal{E} \, d\eta = g(x,y),$$

(A24a)

$$Q(x,y) = \frac{2}{\sqrt{3}} g(x,y) + \frac{2}{\pi\sqrt{3}} \int_{-\infty}^{\infty} \int_{-\infty}^{\infty} g(\mathcal{E},\eta) \frac{(\mathcal{E}-x)^2 - 3(y-\eta)^2}{\left[(\mathcal{E}-x)^2 + 3(y-\eta)^2\right]^2} \, d\mathcal{E} \, d\eta ,$$

(A25b)

which holds under the same conditions as discussed in connection with Eq.(A1).

Closed form solution of more general integral equations with constant coefficients were obtained by Niyogi and Mitra (1973). The result is

$$aQ(x,y) - \frac{1}{2\pi} \int_{-\infty}^{\infty} \int_{-\infty}^{\infty} Q(\mathcal{E}, \eta) \frac{(\mathcal{E} - x)^2 - (\eta - y)^2}{\left[(\mathcal{E} - x)^2 + (\eta - y)^2\right]^2} \, d\mathcal{E} \, d\eta = g(x,y),$$

(A25a)

$$Q(x,y) \;=\; \frac{2}{\sqrt{4a^2-1}} \; g(x,y)$$

$$+ \;\frac{2}{\pi\sqrt{4a^2-1}} \int_{-\infty}^{\infty}\!\!\int_{-\infty}^{\infty} \underset{\odot}{} \; g(\xi,\eta)\;\frac{(2a-1)(\xi-x)^2-(2a+1)(\eta-y)^2}{\left[(2a-1)(\xi-x)^2+(2a+1)(\eta-y)^2\right]^2}\; d\xi d\eta,$$

$$(A25b)$$

for the parameter $\;a > \dfrac{1}{2}\;$, and under similar assumptions as above.

A2. Exact solution of a three-dimensional linear singular integral
equation

The procedure discussed in section A1 may be readily extended
to the three-dimensional case, which was carried out by A.K. Niyogi
(1976). We consider the following linear three-dimensional singular
integral equation with dipole singularity of the kernel:

$$AQ \;\equiv\; a\;Q(x_1,\,x_2,\,x_3)\;-$$

$$-\;\frac{1}{(2\pi)^{3/2}} \int_{-\infty}^{\infty}\!\!\int_{-\infty}^{\infty}\!\!\int_{-\infty}^{\infty} \underset{\odot}{} \; \frac{f(\theta)}{r^3}\; Q(z_1,z_2,z_3)dz_1\; dz_2\; dz_3 \;=\; g(x_1,\,x_2,\,x_3),$$

$$(A26a)$$

where $\;Q(x_1,\,x_2,\,x_3)\;$ is the unknown function and $\;g(x_1,\,x_2,\,x_3)\;$ is
known, and the characteristic $\;f(\theta)\;$ is given by

$$f(\theta) \;=\; \tfrac{1}{2}\,(3\cos^2\theta - 1).\qquad\qquad (A26b)$$

Here, $r$ denotes the distance of $\;P_1(z_1,\,z_2,\,z_3)\;$ from
$P_0(x_1,\,x_2,\,x_3)$ in the Eucledian space $\;E_3\;$ and $\theta$- is the angle
made by the line $P_1\,P_0$ with the $x_1$-axis. The quantity $a$ is a

suitable parameter. The singularity in the kernel at $z_1 = x_1$,
$z_2 = x_2$, $z_3 = x_3$ has been excluded by a small sphere and the
limit is taken as the radius of the sphere approaches zero. For
solving Eq.(A26a), the following assumptions are made:

(i)     In any bounded part of the space $E_3$, the density

$$Q(x_1, x_2, x_3) \in \text{Lip. } \alpha \text{ , } \alpha > 0 \text{ ,}$$

(ii)    At infinity,

$$Q(x_1, x_2, x_3) = O(R^{-k}), \quad k > 0, \quad R^2 = x_1^2 + x_2^2 + x_3^2 .$$

Noting that the characteristic $f(\theta)$ is bounded and continuous in
$\theta$, the necessary and sufficient condition for the existence of the
singular integral

$$\frac{1}{(2\pi)^{3/2}} \int \int_{E_3} \int \frac{f(\theta)}{r^3} Q(z_1, z_2, z_3) \, dz_1 \, dz_2 \, d z_3 \qquad (A27a)$$

is

$$\int_S f(\theta) \, dS = 0, \qquad (A27b)$$

where S is the unit sphere over which $\theta$ - moves (Mikhlin 1965),
which holds in the present case. The existence of a Fourier transform
of singular kernels like (A27a) was established by Calderon and
Zygmund (1952), so that the operator A may be represented, as in
the two-dimensional case, by

$$A Q \equiv F^{-1} \left[ \Phi \, F \, Q \right] = g(\bar{x}), \qquad (A28)$$

where F denotes a Fourier transform in $E_3$, defined by

$$FK \equiv \lim_{\substack{\epsilon \to 0 \\ N \to \infty}} \frac{1}{(2\pi)^{3/2}} \int_{\epsilon < |\bar{x}| < N} e^{-i(\bar{x}\cdot\bar{z})} K(\bar{x}) \, d\bar{x} \qquad (A29)$$

and $\bar{\Phi}$ is the symbol of the integral operator (Mikhlin 1965) defined by

$$\bar{\Phi}(\theta) = a - F\left[\frac{f(\theta)}{r^3}\right]. \qquad (A30)$$

We assume $g(\bar{x}) \in L_2(E_3)$ and in addition that everywhere on the sphere $S$, except on a set of measure zero, the symbol $\bar{\Phi}(\theta)$ is finite and $\inf|\bar{\Phi}(\theta)| > 0$. Then according to Mikhlin (1965) the solution of Eq.(A26a) may be put to the form

$$Q = F^{-1}\left\{\left[\bar{\Phi}\right]^{-1} F g\right\}. \qquad (A31)$$

Noting that the kernel $K(\bar{x})$ of Eq.(A26a) may be rewritten as

$$K(\bar{x}) = \frac{1}{2}\frac{\partial^2}{\partial x_1^2}\left(\frac{1}{R}\right), \qquad (A32)$$

the Fourier transform of the kernel may be expressed in Cartesian coordinates as

$$F\left[\frac{f(\theta)}{r^3}\right] = FK$$

$$= \frac{1}{2(2\pi)^{3/2}}\int_{-\infty}^{\infty}\int_{-\infty}^{\infty}\int_{\bigodot-\infty}^{\infty}\left[e^{-ix_1 z_1}\frac{\partial^2}{\partial x_1^2}\left(\frac{1}{R}\right)dx_1\right] e^{-i(x_2 z_2 + x_3 z_3)} dx_2 \, dx_3.$$

$$(A33)$$

As shown in details by A.K. Niyogi (1976), the above integral may be evaluated in closed form, and the value of the symbol $\Phi$ is obtained as

$$\Phi = a' + \frac{1}{\sqrt{2\pi}} \cos^2\theta_1 = a' + \frac{1}{\sqrt{2\pi}} \frac{z_1^2}{\rho^2} \quad , \qquad (A34a)$$

where

$$a' = a - \frac{1}{\sqrt{2\pi}} \, , \qquad \left| \bar{z} \right| = \rho \, , \qquad \cos\theta_1 = z_1 / \rho \, . $$

$$(A34b)$$

In order to simplify Eq.(A31) by means of Faltung theorem, the inverse Fourier transform $\Phi^{-1}$ is needed, which is found as (A.K.Niyogi 1976)

$$F^{-1}\left[ \Phi \right]^{-1} = \lim_{\varepsilon \to 0} \frac{4}{(b^2 + 1)(\varepsilon^2 + R^2)^2}$$

$$+ \frac{\pi}{c(b^2 + 1)^2} \cdot \frac{2c^2 x_1^2 - x_2^2 - x_3^2}{(x_2^2 + x_3^2 + c^2 x_1^2)^{5/2}} \, , \qquad (A35)$$

where
$$c^2 = b^2 \Big/ (b^2 + 1), \qquad b^2 = \sqrt{2\pi} \, a' \, . $$

It is to be noted that the inverse Fourier transform does not exist in the usual sense and the Abel-Poisson method of summability was used in obtaining Eq.(A35).

Using convolution theorem, the unique solution of Eq.(A26a) is obtained from Eqs. (A31, A35) as

$$Q(x_1, x_2, x_3) = \frac{1}{(2\pi)^{3/2}} \left\{ \lim_{\varepsilon \to 0} \int_{E_3} \frac{4}{(b^2+1)(\varepsilon^2 + r^2)^2} \, g(\bar{z}) \, d\bar{z} \right.$$

$$+ \frac{\pi}{(2\pi)^{3/2} \, c(b^2+1)^2} \int_{E_3} \frac{2c^2(x_1 - z_1)^2 - (x_2 - z_2)^2 - (x_3 - z_3)^2}{\left[ (x_2 - z_2)^2 + (x_3 - z_3)^2 + c^2(x_1 - z_1)^2 \right]^{5/2}} \, g(\bar{z}) d\bar{z},$$

$$(A36)$$

where

$$r^2 = (x_1 - z_1)^2 + (x_2 - z_2)^2 + (x_3 - z_3)^2.$$

Using well-known (Neri 1971) behaviour of the convolution of the Poisson-kernel

$$K_\varepsilon = \frac{1}{\lambda} \frac{\varepsilon}{(\varepsilon^2 + |\bar{x}|^2)^{\frac{n+1}{2}}} \quad ,$$

the first integral in Eq.(A36) may be evaluated and we finally obtain

$$Q(x_1, x_2, x_3) = \frac{1}{a} g(x_1, x_2, x_3)$$

$$+ \frac{1}{2(2\pi)^{3/2}} \cdot \frac{1}{ca^2} \int_{E_3} \frac{2c^2(x_1 - z_1)^2 - (x_2 - z_2)^2 - (x_3 - z_3)^2}{\left[ (x_2 - z_2)^2 + (x_3 - z_3)^2 + c^2(x_1 - z_1)^2 \right]^{5/2}} g(\bar{z}) d\bar{z},$$

$$(A37)$$

as the desired closed form exact solution of Eq.(A26).

# References

Ahlberg, J.H., Nilson, E.N., Walsh, J.L. (1967): The theory of splines and their applications. New York : Academic Press.

Ashley, H. and Landahl, M. (1965): Aerodynamics of Wings and Bodies. Addison-Wesley Inc., U.S.A.

Bailey, F.R. and Steger, J.L. (1972): Relaxation techniques for three-dimensional transonic flow about wings. AIAA Paper No. 72-189, U.S.A.

Bailey, F.R. (1975): On the computation of two- and three-dimensional steady transonic flows by relaxation methods. pp.1-77. In : Progress in Numerical Fluid Dynamics, Ed. Wirz, H.J. Lecture Notes in Physics, vol.41, Springer Verlag, Berlin-Heidelberg - New York.

Basu, S. and Niyogi, P. (1980): Transonic airfoil design at zero incidence. Indian Society for Theoretical and Applied Mechanics (ISTAM) Silver Jubilee Congress, Kharagpur, pp.52-54.

Bauer, F., Garabedian, P. and Korn, D.G. (1972): A Theory of supercritical wing sections, with computer programs and examples. Lecture Notes in Economics and Mathematical Systems, vol.66. Springer Verlag, Berlin/Heidelberg/New York.

Bauer, F., Garabedian, P.R., Korn, D. and Jameson, A. (1975): Supercritical Wing Section II. Lecture Notes in Economics and Mathematical Systems No.108. Springer Verlag, New York.

Baurdoux, H.I. and Boerstoel, J.W. (1968): Symmetric transonic potential flows around quasi-elliptical aerofoil sections. NLR TR 67007 U, Netherlands.

Bers, L. (1954): Existence and uniqueness of subsonic flow past a given profile. Comm. Pure and Appl. Maths., 7, 441-504.

Bers, L. (1958): Mathematical Aspects of Subsonic and Transonic Gasdynamics. John Wiley and Sons, INC., New York.

Bitsadze, A.V. (1964): Equations of Mixed Type. Translation from Russian. Macmillan, New York.

Boerstoel, J.W. (1967): A survey of symmetrical transonic potential flow around quasi-elliptical aerofoil sections. NLR TR T.136, Netherlands.

Calderon, A.P. and Zygmund, A. (1952): On the existence of certain singular integrals. Acta Math. 88, 85-139.

Chakraborty, S.K. (1974): An analytical and numerical study of plane transonic profile flow at zero and non-zero incidence. Ph.D.(Sc.) thesis, Jadavpur University, Calcutta.

Chakraborty, S.K. (1975): Shock-free transonic flow past lifting airfoils. AIAA Journal, 13, 1094-1097.

Chakraborty, S.K. and Niyogi, P. (1977): Integral equation formulation for transonic lifting profiles. AIAA Journal, 15, 1816-1817.

Chakraborty, S.K. (1978): Integral Equation Formulation for Transonic Flow past Lifting Wings. AIAA Journal, 16, 1015-1016.

Cole, J.D. (1969): Twenty Years of Transonic Flow. Boeing Scientific Research Laboratories D1-82-0878, U.S.A.

Cole, J.D. and Messiter, A.F. (1957): Expansion procedures and similarity laws for transonic flow. ZAMP, 8, 1-25.

Cole, J.D. (1975): Modern Developments in Transonic Flow. SIAM J. Appl. Math. 29, 763-786.

Courant, R. and Hilbert, D. (1962): Methods of Mathematical Physics, Vol.II, John Wiley and Sons., New York, INC.

Epstein, B. (1962): Partial Differential Equations, McGraw-Hill Inc., New York.

Ferrari, C. and Tricomi, G. (1968): Transonic Aerodynamics. English translation. Academic Press, New York.

Frohn, A. (1974): Lösung der nichtlinearen Integralgleichung der schallnahen Strömung. ZAMM, 54, T116-T117.

Frohn, A. (1976): Problems and results of the integral equation method for transonic flows. In : Oswatitsch, K. and Rues, D. (Eds.) Symposium Transonicum II., pp.191-196. Springer-Verlag, Berlin Heidelberg, New York 1976.

Garabedian, P.R. and Korn, D.G. (1971): Analysis of transonic airfoils. Comm. Pure Appl. Math. 24, 841-851.

Guderly, G. (1946): Die Ursache für das Auftreten von Verdichtungsstossen in gemischten Unter-Überschallströmungen. M.O.S. (A) Volkenrode, Rd T 110.

Gullstrand, T.R. (1951): The flow over symmetrical aerofoils without incidence in the lower transonic range. KTH AERO TN 20.

Gullstrand, T.R. (1952): The flow over symmetrical aerofoils without incidence at sonic speed. KTH AERO TN 24.

Gullstrand, T.R. (1952): The flow over two-dimensional aerofoils at incidence in the transonic speed range. KTH AERO TN 27.

Gullstrand, T. (1953): Transonic flow past two-dimensional aerofoils. Zeit. für Flugwissenschaften, 1,

Gretler, W. (1965): Neuere Methode zur Berechnung der ebenen Unterschallströmung an dunnen Profilen bei kleinen Anstellwinkeln. Acta Mechanica, 1/2, 109-134.

Gretler, W. (1971): Anwendung direkter und indirekter Methoden in der Theorie der Unterschallströmungen. In : E.Leiter and J.Zierep (Eds.): Übersichtsbeiträge zur Gasdynamik, 95-116. Springer Verlag Vienna/New York.

Haldar, K. (1972): On some problems of high-speed gasdynamics. Ph.D. (Sc.) thesis, Jadavpur University, Calcutta.

Hansen, H. (1975): Entwurf von Tragflügelprofilen für schallnahe Anströmung nach der Integralmethode. Internal Report of the Institute for Aerodynamics of DFVLR, IB 151-75/4, Braunschweig.

Hansen, H. (1976a): Entwurf von Tragflugelprofilen für schallnahe Anströmung nach der Integralmethode. Zeit. für Flugwiss., 24, 340-342.

Hansen, H. (1976b): Airfoil design for a prescribed velocity distribution in transonic flow by an integral method. Symposium Transonicum II, Eds. K. Oswatitsch and D. Rues, Berlin Springer Verlag, 183.

Hays, W.D. (1966): La seconde approximation pour les écoulements transonique non visqueux. Jour. de Mechanique, 5, 163-206.

Heaslet, M.A. and Spreiter, J.R. (1957): Three dimensional transonic flow applied to slender wings and bodies. NACA Rep. 1318, U.S.A.

Hess, J.L. (1969): Calculation of potential flow about arbitrary three-dimensional lifting bodies. Douglas Aircraft Company, Report No.MDC-j0545, U.S.A.

Jameson, A. (1974): Iterative solution of transonic flows over airfoils and wings, including flow at Mach 1. Comm. Pure Appl. Math. 27, 283-309.

Jameson, A.        : Transonic flow calculations for airfoils and bodies of revolution. Grumman Aerospace Corporation, U.S.A. Aerodynamics Report 390-71-1, undated.

Jones, A.F. (1971): Measured pressure distribution for a swept wing at transonic speeds. Royal Aircraft Establishment, Tech. Memo. Aero 1282, England.

Keune, F. (1952): Low aspect ratio wings with small thickness at zero lift in subsonic and supersonic flow. KTH AERO TN 21.

Keune, F. and Oswatitsch, K. (1953): Nicht angestellte Körper kleiner Spannweite in Unter - und Überschallströmung. Zeit.f. Flugwissenschaften, 1, p. 137-145.

Keune, F. and Oswatitsch, K. (1955): An integral equation theory for the transonic flow around slender bodies of revolution at zero incidence. KTH AERO TN 37.

Klunker, E.B. (1971): Contributions to the methods for calculating
the flow about thin lifting wings at transonic speeds - Analytical
expressions for the far field. NASA TND - 6530, U.S.A.

Kluwick, A. and Oswatitsch, K. (1974): Unstetigkeitseigenschaften
der Integralgleichung für stationäre räumliche schallnahe Strömungen.
Ommagio a Carlo Ferrai. Librareria Editrice Universitaria Levrotto
and Bella, 415-429, Torino.

Kundu, A. (1972): Iterative solution of the integral equation of
Oswatitsch. Master's Thesis, P.G. Diploma course in Computer Science,
Jadavpur University, Calcutta.

Leathem, J.G. (1960): Volume and Surface Integrals Used in Physics.
Hafner Publishing Co., New York.

Lock, R.C. (1970): Test cases for numerical methods in two-dimensional transonic flow. AGARD Report No.575.

Manwell, A.R. (1971): The Hodograph Equations. An Introduction to
the Mathematical Theory of Plane Transonic Flow. Oliver and Boyd,
Edinburgh.

Mikhlin, S.G. (1965): Multi-dimensional singular integrals and
integral equations. English translation. Pergamon Press, London.

Mitra, R. (1976): Approximate transonic profile flow  with shocks.
Acta Mechanica, $\underline{25}$, 1-12.

Morawetz, C.S. (1956): On the non-existence of continuous transonic
flows past profiles I. Comm. Pure Appl. Math., $\underline{9}$, 45-68.

Murman, E.M. and Cole, J.D. (1971): Calculation of plane steady
transonic flows. AIAA Journal, $\underline{9}$, 114-121.

Murman, E.M. (1974): Analysis of embedded shock waves calculated by
relaxation methods. AIAA Journal, $\underline{12}$, 626-632.

Neri, U. (1971): Singular integrals. Lecture Notes in Mathematics,
Chapter I. Springer Verlag, Berlin-Heidelberg-New York.

Nieuwland, G.Y. (1967): Transonic potential flow around a family
of quasi-elliptical aerofoil sections. NLR Report TR. T172, Amsterdam.

Nixon, D. (1974): High Subsonic Flow Past a Steady Finite Wing.
QMC EP-1015, Queen Mary College, London.

Nixon, D. and Hancock, G.J. (1974): High Subsonic Flow past a Steady
two dimensional  Airfoil. ARC CP 1280, London.

Nixon, D. and Patel, J. (1975): The evaluation of an Integral
Equation Method for two dimensional shock-free flows. Aero.
Quart., $\underline{26}$, 59-70.

Nixon, D. (1975a): Extended Integral Equation Method for Transonic
Flows. AIAA Journal, $\underline{13}$, 934-936.

Nixon, D. (1975b): A comparison of Two Integral Equation Methods
for High Subsonic Lifting Flows. Aero. Quart., $\underline{26}$, 56-58.

Nixon, D. (1976a): An Extended Integral Equation Method for the steady transonic flow past a two-dimensional aerofoil. In : Computational Methods and Problems in Aeronautical Fluid Dynamics. Ed. Hewitt, B.L. et al., Academic Press, London, 270-289.

Nixon, D. (1976b): Reply to Author to H.Nørstrud. AIAA Journal, 14, 828-830.

Nixon, D. and Hancock, G.J. (1976): Integral equation methods - a reappraisal. In : Symposium Transonicum II, Ed. Oswatitsch, K. and Rues, D.pp. 174-182, Springer-Verlag, Berlin Heidelberg New York.

Nixon, D. (1977): Calculation of Transonic Flows using an Extended Integral Equation Method. AIAA Journal, 15, 295-296.

Nixon, D. (1978): Direct Numerical Solution of the Transonic Perturbation Integral Equation for Lifting and Nonlifting Flows. NASA TM 78518, U.S.A.

Nixon, D. (1979): The Transonic Integral Equation Method with Curved Shock. Acta Mechanica, 32, 141-151.

Niyogi, P. (1969): Exact solution of a two dimensional singular integral equation of interest in plane transonic flow. Indian J. of Math. Mech., Part II Special Issue, 109-112, Calcutta.

Niyogi, P. (1973): Exact solution of a linear two dimensional singular integral equation. ZAMM, 53, 413-415.

Niyogi, P. and Mitra, R. (1973): Approximate shock-free transonic solution for a symmetric profile at zero incidence. AIAA Journal, 11, 751-754.

Niyogi, P. (1976): Shock-free transonic flow past symmetric profiles at zero incidence. Lecture delivered at the Symposium on Continuum Mechanics, Centre of Advanced Study in Applied Mathematics, Calcutta University, March 1972. Bull. Cal. Math. Society, 68, 77-86.

Niyogi, A.K. (1976): Exact solution of a three-dimensional singular integral equation. ZAMM, 56, 173-175.

Niyogi, P. (1977): Inviscid Gasdynamics. The Macmillan Company of India, New Delhi.

Niyogi, P. (1978a): Transonic integral equation formulation for lifting profiles and wings. AIAA Journal, 16, 92-94.

Niyogi, P. (1978b): On existence and uniqueness of plane transonic flows. Lecture delivered at the 23rd Congress of Indian Soc.Theor. and Appl. Mech., R.E.College, Warangal, Conference abstracts.

Niyogi, P. and Sen, R. (1978): Approximate transonic profile flow at incidence with shock. Acta Mechanica, 30, 65-77.

Niyogi, P. and Chakraborty, S.K. (1979): Iterative computation of transonic shock-free profile flow at zero incidence. Acta Mechanica, 31, 173-184.

Niyogi, P. (1979): Trailing Edge Lift Distribution of Small Aspect Ratio Wings. Jadavpur University, Calcutta.

Niyogi, P. and Das, T.K. (1979): Direct computation of transonic solution for Nieuwland aerofoils. Acta Mechanica, 34, 285-289.

Niyogi, P. and Sen, S. (1979, Unpublished): Symmetric transonic shock-free airfoil design. To appear in the Bull. Cal. Math. Society, Calcutta.

Niyogi, P. (1980a): Transonic flow past thin wings. Proceedings of Indian Society for Theoretical and Applied Mechanics Conference, International Symposium on Non-linear Continuum Mechanics, pp.14-16. Kharagpur

Niyogi, P. (1980b): Recent developments in integral equation method in transonic flow. Proc. Indian Acad. Sci. (Engg. Sci.), 3, 143-167.

Niyogi, P. (1981): Transonic flow past thin wings. Lecture delivered at the First Asian Congress of Fluid Mechanics, Bangalore, December 8-13, 1980. Proc. Indian Acad. Sci. (Engg. Sci.), 4, 347-361.

Niyogi, P.         : Existence and uniqueness of shock-free transonic solution for symmetrical thin wings at zero incidence. To appear in Acta Mechanica.

Niyogi, P. and Das, T.K. (1981): Plane transonic solution with shock by direct iteration. Acta Mechanica 38, 169-181.

Niyogi, P. and Ray, A.K. (1981): A new iteration scheme for transonic aerodynamic problems. Indian Society for Theoretical and Applied Mechanics Congress, Coimbatore, December 1981. Conference Abstracts.

Nørstrud, H. (1968): Numerische Lösungen von schallnahen Strömungen um ebene Profile. Dissertation, TH Wien.

Nørstrud, H. (1970): Numerische Lösungen für schallnahe Strömungen um ebene Profile. Zeits. für Flugwiss. 18, 149-157.

Nørstrud, H. (1971): Three dimensional nonlinear flow over finite symmetrical wings of arbitrary planform. Acta Mechanica, 11, 299-312.

Nørstrud, H. (1973a): High Speed Flow Past Wings. NASA CR-2246.

Nørstrud, H. (1973b): The transonic aerofoil problem with embedded shocks. Aero. Quarterly, 24, 129-139.

Nørstrud, H. (1976): Comment on, 'Extended Integral Equation Method for Transonic Flows', AIAA Journal, 14, 826-828.

Nørstrud, H. (1973c): Transonic flow past lifting wings, AIAA Journal, 11, 754-757.

Ogana, W. and Spreiter, J.R. (1977): Derivation of an integral equation for transonic flows. AIAA Journal, 15, 444-446.

Ogana, W. (1979): Derivation of an integral equation for three-dimensional transonic flows. AIAA Journal, 17, 305-307.

Oswatitsch, K. (1950): Die Geschwindigkeitsverteilung an symmetrischen Profilen beim Auftreten lokaler Überschallgebiete. Acta Physica Austriaca, 4, 228-271. Also, ZAMM, 30, 17-24. English translation: The velocity distribution on symmetric Profiles with local supersonic regions, pp. 150-187, In : Contributions to the Developments of Gasdynamics, Eds. W. Schneider and M.Platzer. Fiedr. Vieweg and Sohn, Braunschweig/Wiesbaden (1980).

Oswatitsch, K. and Keune, F. (1954): Ein Äquivalenssatz für nicht angestellte Flügel kleiner Spannweite in schallnaher Strömung. Zeit. für Flugwissenschaften, 3, p. 29.

Oswatitsch, K. (1956): Gasdynamics. English translation. Academic Press, New York, N.Y.

Oswatitsch, K. (1959): Physikalische Grundlagen der Strömungslehre. Handbuch der Physik, Vol.VIII/1, Ed. S.Flügge, pp. 1-124. Springer Verlag, Berlin/Göttingen/Heidelberg.

Oswatitsch, K. (1960): Similarity and Equivalence in Compressible Flow. Advances Appl. Mech., Vol.VI, pp. 153-271. Academic Press, New York.

Oswatitsch, K. and Zierep, J. (1962): Stationäre ebene schallnahe Strömung. DVL - Bericht Nr. 189. Proz-Wahn, West Germany.

Oswatitsch, K. (1976): Grundlagen der Gasdynamik. Springer Verlag, Vienna.

Oswatitsch, K. (1977): Spezialgebiete der Gasdynamik. Springer Verlag, Vienna.

Pearcy, H.H. (1962): The aerodynamic design of section shapes for swept wings. Adv. Aero. Sci., vol.3, 277-322.

Rall, L.B. (1969): Computational solution of nonlinear operator equations. Chapter 2. John Wiley and Sons, Inc., New York.

Roberts, G.E. and Kauffman, H. (1966): Tables of Laplace Transforms. W.B. Saunders Company, New York, p. 13.

Rubbert, P.E. and Landahl, M.T. (1967): Solution of transonic airfoil problem through parametric differentiation. AIAA Journal, 5, 470-479.

Schmitt von Schubert, B. (1980): Integralgleichung für das Geschwindigkeitsfeld drehungsfreier Strömungen um stumpfe Körper. DFVLR Forschungsbericht 80-25, Koln.

Schneider, W. (1978): Mathematische Methoden der Strömungsmechanik. Verlag Vieweg, Braunschweig, W.Germany.

Schubert, H. and Schleiff, M. (1969): Über zwei Randwert-probleme des inhomogenen Systems der Cauchy-Riemannschen Differential-gleichungen mit einer Anwendung auf ein Problem der stationären schallnahen Strömung. ZAMM, 49, 621-630.

Schubert, H. (1973): Bemerkung zu einer zweidimensionalen singulären Integralgleichung. ZAMM 53, 415-416.

Sen (née Mitra), R. (1976): Study in transonic profile flow. Ph.D.(Sc.) thesis, Jadavpur University, Calcutta.

Serrin, J. (1959): Mathematical Foundations of Fluid Mechanics, Handbuch der Physik, Vol.VIII/1, Ed. S.Flügge, pp. 125. Springer Verlag, Berlin/Göttingen/Heidelberg.

Spee, B.M. and Uijlenhoet, R. (1968): Experimental verification of shock-free transonic flow around quasi-elliptical aerofoil sections. NLR Report MP. 68003 U, Amsterdam.

Spreiter, J.R. (1954): On alternative forms for the basic equations of transonic flow theory. J.Aero. Sci., 21, 70-72. Errata : J. Aero soc., 21, 360.

Spreiter, J.R. and Alksne, A.Y. (1955): Theoretical Prediction of Pressure Distributions on Nonlifting Airfoils at High Subsonic speeds. NACA Rep. 1217, U.S.A.

Steger, J.L. and Lomax, H. (1972): Transonic flows about two dimensional airfoils by relaxation procedures. AIAA Journal, 10, 49-54.

Tricomi, F.G. (1923): Sulle equazioni lineari alle derivate parziali di secondo ordine, di tipo misto. Rendiconti, Atti dell' Academia Nazionale dei Lincei, Series 5, 14, 134-247.

Tricomi, F. (1926): Formula dell' inversione dell' ordine d'une integrazioni doppio conastrico. Rend. Acc. Naz. Lincei, 3. Ser 6a, fasc. 9, 535-539.

Tricomi, F.G. (1957): Integral equations, Chap. 4. Interscience, New York.

Truckenbrodt, E. (1951): Die Berechnungder Profil bei Vorgegebener Geschwindigkeitsverteilung. Ing.-Arch. 19, 365-377.

Vekua, I.N. (1963): Veralgemeinerte analytische Funktionen. Berlin. Also : Generalized Analytic Functions. Oxford, Pergamon Press 1962.

Weber, J. (1957): The calculation of the pressure distribution on the surface of thick cambered wings and the design of wings with given pressure distribution. ARC R and M 3026, London.

Woodward, F.A. (1968): Analysis and design of Wing-body combinations at subsonic and supersonic speeds. Journal of Aircraft, 5, 528-534.

Von Wolfersdorf, L. (1962): Zur Lösung der einfachsten zweidimensionalen singulären Integralgleichung. Rend. del Seminario Mathematico dell' Universita e del Politechnico di Torino, 22, 215-226.

Yakovlev, M.N. (1964): The solution of systems of nonlinear equations by a method of differentiation with respect to a parameter. USSR Comp. Mathematics and Math. Physics, 14, 198-203.

Yoshida, K. (1960): Lectures on Differential and Integral Equations. Interscience, New York.

Zierep, J. (1962): Die Integralgleichungsmethode zur Berechnung schallnaher Strömungen. In: Sympsium Transonicum, Aachen 1962, Ed. Oswatitsch, K., pp. 92-109. Springer Verlag, Berlin 1964.

Zierep, J. (1966): Theorie der schallnahen und der Hyperschall-strömungen. Verlag G. Braun, Karlsruhe.

Zierep, J. (1976): Vorlesungen über Gasdynamik, 3. Aufl. Verlag G. Braun, Karlsruhe.

# Finite-Difference Techniques for Vectorized Fluid Dynamics Calculations

Editor: D.L. Book
1981. 60 figures. VIII, 226 pages
(Springer Series in Computational Physics)
ISBN 3-540-10482-8

**Contents:** Introduction. – Computational Techniques for Solution of Convective Equations. – Flux-Corrected Transport. – Efficient Time Integration Schemes for Atmosphere and Ocean Models. – A One-Dimensional Lagrangian Code for Nearly Incompressible Flow. – Two-Dimensional Lagrangian Fluid Dynamics Using Triangular Grids. – Solution of Elliptic Equations. – Vectorization of Fluid Codes. – Appendices A-E. – References.

R. Peyret, T.D. Taylor

# Computational Methods for Fluid Flow

1982. Approx. 126 figures. Approx. 320 pages
(Springer Series in Computational Physics)
ISBN 3-540-11147-6

**Contents:** Numerical Approaches: Introduction and General Equations. Finite Difference Methods. Integral and Spectral Methods. Relationship Between Integral, Spectral and Finite Difference Methods. Specialized Methods. – Incompressible Flows: Finite Difference Solutions of the Navier-Stokes Equations. – Finite Element Methods Applied to Incompressible Flow. Spectral Method Solutions for Incompressible Flows. Turbulent Flow Models and Calculations. – Compressible Flows: Inviscid Compressible Flow. Viscous Compressible Flows. – Concluding Remarks.

# Symposium of Numerical and Physical Aspects of Aerodynamic Flows

19–21 January 1981
California State University, Long Beach, California
Editor: T. Cebeci
1982. Approx. 346 figures. Approx.  pages
ISBN 3-540-11044-5

This symposium treated numerical and physical aspects of aerodynamical flows. The contributions have been updated for publication and cover Numerical Fluid Dynamics, Interactive Steady Boundary Layers, Singularities in Unsteady Boundary Layers, Transonic Flows and Experimental Fluid Dynamics. Each section begins with a critical review and introduces the reader to the papers which follow. It is hoped that this volume provides a good overview of current knowledge, helps to set priorities for future developments and will due to the careful editing keep its value over the years.

D.P. Telionis

# Unsteady Viscous Flows

1981. 132 figures. XXIII, 408 pages
(Springer Series in Computational Physics)
ISBN 3-540-10481-X

**Contents:** Basic Concepts. – Numerical Analysis. – Impulsive Motion. – Oscillations with Zero Mean. – Oscillating Flows with Non-Vanishing Mean. – Unsteady Turbulent Flows. – Unsteady Separation. – References.

F. Thomasset

# Implementation of Finite Element Methods for Navier-Stokes Equations

1981. 86 figures. VI, 162 pages
(Springer Series in Computational Physics)
ISBN 3-540-10771-1

**Contents:** Introduction. Notations. – Elliptic Equations of Order 2: Some Standard Finite Element Methods. – Upwind Finite Element Schemes. – Numerical Solution of Stokes Equations. – Navier-Stokes Equations: Accuracy Assessments and Numerical Results. – Computational Problems and Bookkeeping. – Appendix 1: The Patch Test of the $P1$ Nonconforming Triangle: Sketchy Proof of Convergence. – Appendix 2: Numerical Illustration. – Appendix 3: The Zero divergence Basis for 2-D $P1$ Nonconforming Elements. – Three Dimensional Case. – References.

Springer-Verlag
Berlin
Heidelberg
New York

# Lecture Notes in Physics

# Selected Issues from
# Lecture Notes in Mathematics